UNIVERSE LOST

Reclaiming A Christian World View

UNIVERSE LOST

Reclaiming A Christian World View

Stuart Cook

Copyright © 1992
College Press Publishing Company

Printed and Bound in the
United States of America
All Rights Reserved

Cover Photo: Ed Pritchard/Tony Stone Worldwide

Library of Congress Catalog Number: 91-77277
International Standard Book Number: 0-89900-404-0

DEDICATION

A man named Dunning made me think deeply;
A man named Lackey made me think more accurately;
A man named Brinsmead made me think about important things;
But God made it possible for me to think real thoughts.

CONTENTS

	INTRODUCTION	9
1	The *Big* View of History	15
2	So What?	29
3	"And He Went Forth, Not Knowing..."	35
4	Let Them Have Dominion..."	39
5	"They Did Not Like To Retain God..."	45
6	How Naturalism Became Unnatural	51
7	It's Nothing New, But Is It True?	69
8	"The Creature Rather Than the Creator"	87
9	"As A Man Thinketh in His Heart..."	97
10	Getting to the Heart of the Matter	111
11	What Do You Mean, "The Word of God"?	117
12	Who's Responsible?	135
13	Man — An Active Ingredient	143
14	But I Need Help!	151
15	Clearing Away the Mists	165
16	Doing the Truth	175
17	Where's It All Going to End?	189
	BIBLIOGRAPHY	193

Introduction

Should a professional theologian or philosopher consider reading this book, let me state that the epistemological theses herein presented conclude that problematic cosmological formulations tend to be highly conducive to cognitive dysfunctionality.

For the benefit of the rest of us, the bottom line of this book is that if we don't believe straight, we can't think straight. We need some straight talk about straight thinking. Most of us thoroughly enjoy expressing our opinions, but our enthusiasm would doubtless dissipate considerably were we asked instead, "Tell me, *how* did you decide to believe that, and *why* do you reason as you do?" About the only experience more threatening than having the very foundations of our beliefs challenged, is to also have our rationality put in doubt.

The world is in a growing state of turmoil, and the central issue, believe it or not, hinges on the question of what constitutes reality! Because this is so, it's vital that we bring two things into agreement: the facts of reality — what really and truly exists and does not exist and what it is that causes things to happen as they do — and our personal understanding of these facts. Reality won't alter to suit our understanding. We must conform our understanding to reality, not the reverse. Misguided sincerity can be *fatal*.

The writer, being a Christian, draws his own personal interpretation of reality from the Bible, which he takes very seriously, and therefore the ideas in this book are expressed from that perspective. It's assumed that the majority of readers will also classify themselves as Christians. That doesn't mean that we'll be in agreement, however. It isn't easy to admit, true though it is, but there are people who are neither ignorant, stupid or dishonest and yet, categorically disagree with me!

Of course, this is not to infer that in spite of our differences we are all equally right, just as long as we are normally intelligent, sincere and of course harmless.[1] Honestly believing something doesn't thereby make it true, and the outcome of some of these differences is legitimate cause for concern. The growing influence of mysticism in Christian circles, the theme of this book, makes for an explosive and controversial topic. The implications of this phenomenon make it practical.

At this point you're going to be put "on the spot": Can you, clearly and accurately, explain the difference between things that are *supernatural* and things that are *mystical*? In all probability, you've never tried and can't see any good reason to start now. It is important, and not simply for the sake of defining words. These two words are sometimes used as if they were synonyms, sometimes as contrasting, and in an increasing manner, today, as if mysticism were a subdivision of the broader category of the supernatural. In the pages ahead we'll define, step-by-step, two vastly different and incompatible ways of thinking, of explaining reality, and to these we'll give the names "supernature" and "mystical." These differences are showing themselves in a major way both outside and within the church, and are producing two very different ways of acting, as well as evaluating our experiences and making decisions.

An example: Picture yourself holding a conversation with a new acquaintance. The two of you have just met at a dinner party and have plunged into interesting and lively discussions on all sorts of topics. Opinions are exchanged on social problems, the Middle-East crisis and our current economic difficulties. Both of you express your views and are agreed that the world situation leaves plenty of room for improvement.

Your new-found friend, however, does seem considerably more optimistic about the future of society than you yourself tend to be, and has even appeared to be hinting during the conversation that he might possess some kind of special inside information. Finally, it comes out when he says something like this:

"Yes, Planet is in danger, but we're making progress. The threat of global nuclear holocaust and ecological destruction is still very real, but there's also a growing awareness that Planet can be saved if humanity achieves spiritual oneness. God is in everyone, if only they knew it. In fact, world-wide, more people are becoming conscious of the Universal Mind. We will succeed, I'm sure, but traditional religion still poses a threat to survival. Some of the leaders are saying that for survival's sake it might be necessary to release the hold-outs from this incarnation if holism is to be achieved. I suppose there always will be a few 'die-hards,' if you'll pardon the pun. You surely agree, don't you?"

"Well," you say to yourself, "there will always be a few nut-cases at least! But if I rightly understand what this guy is saying, he could be dangerous!"

But that other person doesn't think he's a nut-case or dangerous, and in is most likely to be of at least normal sanity, intelligence and education. I use this illustration because it probably seems wild and way-out to the average reader. It might not be quite as uncommon as you think.[2]

Let's use another illustration, an experience which my wife and I had as we were calling in a home to talk about Jesus Christ. "Suppose for the sake of discussion," I asked the man, "suppose I could produce absolute proof, by whatever standard you think is real proof, that Jesus genuinely is the Son of God, was crucified and resurrected, and on that basis calls on you to make Him the Lord of your life. Would you consider doing it?"

The man seemed honestly shocked at my question. "Of course not! I have my own beliefs!" To me this seemed like utter nonsense. How could one have "absolute proof" of one fact and still believe it is rational to hold his "own beliefs" to the contrary? Yet to him it was self-evidently sensible.

Or consider the young woman who became frightened and strongly disapproving when I asked her to explain what she meant by "God," since she claimed to believe in Him.

"Oh, I would never try to define God! It just goes against my whole nature. You don't dare take such a rational approach because all it can lead to is darkness and despair!"[3]

Or what of the people — of whom there are multitudes in some cultures — who, with logical consistency from their way of thinking, will change ideologies when they change environments. On Sunday in the city they are Christians. In the factory with the labor union they are Communists. At home, in the

INTRODUCTION

rural area with their family, they sacrifice and pray to ancestral spirits.[4]

When does something make sense? What is logical and rational? This all depends on what a person holds to be "real," what is believed to truly exist or not exist, and what one thinks it is that causes events to happen as they do. There is obviously nothing irrational or stupid about any belief that agrees with things as they truly *are*. But what truly *is*? There's the rub!

We can't avoid the fact that there are ways of thinking — and then again there are ways of thinking. It is also true that the world's way of thinking has been baptized, given a Christian name and is being inducted as a full member — without ever giving up its old character. And this is bad news!

It's very important to clearly understand what is happening as a result of the confusion between mysticism and supernaturalism. Among the issues which are involved are such matters as how to properly understand the Bible, what it is that makes the cults so popular, the nature and source of the very real power in non-Christian religions, whether or not God approves of Christians making plans and setting goals for the future, why the "New Age" movement and Christianity are incompatible, what miracles are, and a whole lot more. It alters what is meant by the term, "spiritually-minded." Or, if someone is described as having been a "great mystic," does that mean that he was a Godly man in some special kind of way, and if so, in what way? With a list like this, it's obvious that we will be fitting pieces of the puzzle together, but not analyzing each piece in fine detail. The individual details of that puzzle only make sense once we have the larger picture in mind.

Now that we've discussed *where* we're going in our thinking, let's go, but watch out! The next step is a lo-o-ong one!

Notes — Introduction

1. Prov. 14:12; 16:2
2. The thinking represented in this illustration is documented in books such as:

FERGUSON, Marilyn: *The Aquarian Conspiracy: Personal and social transformation in the 1980s*; NY: St. Martin's Press, 1980

CUMBEY, Constance: *The Hidden Dangers of the Rainbow*; Shreveport, LA: Huntington House, Inc. 1983

HUNT, Dave and MCMAHON, T.A.: *The Seduction of Christianity*; Eugene, OR: Harvest House Publishers, 1985

3. This experience occurred in 1970 just a few months prior to my first reading Francis A. Schaeffer's *Escape From Reason*, in which he speaks of the "line of despair." At the time I wondered how a well-educated and apparently intelligent young woman could be so irrational.

SCHAEFFER, Francis A.: *Escape From Reason*; London: Intervarsity Fellowship, 1968

4. This accommodation of the environment continues to be the guiding philosophy in Africa, where group membership and authority figures create the standards to which one must conform, in contrast to abstract, objective principles. Thus a man in a position of power may make decisions which to a conceptually-oriented mind are contradictory, inconsistent and even unethical, but are entirely reasonable to African cultures in which the consistency is to be found in the *person*, not in an abstract *idea*.

1

The Big View Of History

AHEAD OF TIME!

"In the Beginning" isn't quite early enough. Let's go back *before* there was a beginning to everything that we know or can see — before there was a "Universe."[1] The fact of the matter is that the Universe isn't universal! Before all of this came into existence — this whole vast and wonderful cosmic panorama of time and space and stars and planets and distant galaxies and whatever else is to be found — there was *another* universe, a totally different universe. Let's call it a "universe" even if it seems strange to use a word which should mean "all that there is" in this way.

But where the other universe is to be found can't be described in terms of directions and distances from earth. No one can accurately point out into space and say, "If we could just go enough light years in *that* direction, eventually we'd come to another universe." If that were the case it would not be *another* universe at all, but just a remote section of this one which we occupy.

The other universe is really and truly *Other*. It is completely independent of this universe in which we are living and occupies different space, operates within its own time-framework, and is composed of its own kind of material substances which are not the same as the atoms and molecules found in this universe. Let's call this other universe *Heaven*. That's what its Owner calls it.[2]

Heaven is God's universe. It is a very real place, even more real than this universe since it existed before this one and will continue to exist when this one is gone. And it isn't "real" in some vague and abstract way, like thoughts and emotions are real to us. If we were there we would know just what *real* reality is.

If we could somehow be transported back and back, before the beginning of this universe, we would be in Heaven, because that would be the only place which was. If we could have seen it, the beauty would have been so magnificent that we could never have dreamed of a way to improve anything. After all, GOD is there![3]

God, in all His Glory. God the Heavenly Father — Heaven's Father. God Who is everywhere and yet is someplace specific — in Heaven. Just because He is everywhere in general doesn't mean that He is no-where in particular. Heaven is Home to God. Just think how glorious Heaven is, if it is the dwelling-place of the Infinite, Singular/Plural, Eternal, Ultimate One. No, it's no use, we are not even capable of imagining it. But we know it has to be so.[4]

If we had been there in Heaven "before the beginning," we

would have seen that Heaven was populated. In fact the population was quite substantial — what kind of — people? — would we have seen populating Heaven?

Should we even call them people? They are distinct *personalities*, these Heavenly beings, each one with his/her/its own unique qualities and abilities. We call them angels. There are more than we could count, and they exist in even greater varieties of kinds and sizes and characteristics than the people we are acquainted with.[5] Anyone who has trouble being comfortable with mankind's racial differences here would really be in a fix in Heaven! We would probably have also seen other kinds of beings and been unable to rightly decide whether to call them angels or something else. We might have had to settle on just "living creatures" — created beings.[6] One thing for certain is that Heaven is not empty!

Observe this important point: Only GOD is omnipresent, that is, unlimited by space.[7] Every single one of His created beings is limited in this way and can't be in two places simultaneously. To exist in Heaven angels must have space to exist in. To be active, to do anything, there must be space — three-dimensional? Could be! — to act and move around in.

What about *time*? Does Heaven have time? Certainly not anything which can be measured with our universe's seconds and minutes and years. But if Heaven has action and space, it also has its own time in which these actions take place. At least let's call it Time, for lack of any better term.

Now, what about *substance*, matter, mass — "stuff" of some sort? Would being in Heaven be like floating in the air, being a vaporous ghost, weightless and transparent, like the silly cartoons of floating on a cloud with wings and a harp? No way! How do we know? Because when Jesus Christ went back there to sit at His Father's right hand, He had a *real body* which He took with Him. His resurrected body was immortal and He had super powers such as walking through walls, but it was real. "Touch Me and see," He said. He ate some food to prove that He was solid and tangible. But the body He now had was made of a superior "stuff" to anything this earth is familiar with. That body was very definitely real, and a real body must have a real environment to be in. Right now that Body, with its Infinite Divine Spirit, is in Heaven.[8] What a staggering concept!

We have no way of knowing many details about that wonderful environment, but we can be certain that we would be in the midst of an orderly system, because God is not the Author of confusion.[9]

The nature of Heaven is super, so let's just call that whole universe the *SUPERNATURAL* realm!

"LET THERE BE . . ."

Imagine witnessing God create a universe! While throngs of amazed angels watch, God, acting through the instrumentality of the One Who later would be known as Jesus, His Son, just *speaks* a whole new system into existence.[10]

This new universe, this "heavens and earth" which God is creating, is not being formed "way out there" somewhere, but in a real way all around us, and yet definitely not just as an annex to what already is. From our vantage-point in Heaven, we can see into another realm, made out of other "stuff" which moves around in its own space and time.

There is an incredible amount of power and energy in this new universe, which doesn't surprise us since it comes from the Ultimate Source of power, God.[11] We know that energy only comes from energy, and we see God as He, shall we say, "reconstructs" or "rearranges" some of His power into a different form. As a result there is now a brand-new natural system made out of material which is basically just energy particles — electrons and all kinds of other things — acting and reacting on themselves. Since they are put together in a different way, they can't act on or react to the "stuff" of the supernatural realm. In fact, each system goes on just as if the other one were not even there.

In this new universe we can see that God has made a vast array of objects ranging in size from either extreme, each atom sort of a miniature galaxy, while innumerable clusters of stars move together through space in concert with one another. The whole realm is a beautiful system of interacting and interdependent parts moving with the harmony of an incredibly complicated instrument, because God has made it in such a way as to be, temporarily,[12] self-perpetuating. It is orderly and predictable, to the extent that we are able to learn what the pattern is and how all the astonishingly complex components function.

Since we now have two separate universes simultaneously in existence, we need to name this new one. The other we called the "Supernatural" realm. What could be more natural than to call this one the *NATURAL UNIVERSE?*

Pausing for a moment in our imaginary journey back to this universe's creation, let's compare these two realms with one another to clarify our thinking if we can. How shall we picture it to clear up our world-view? Various writers have fictionalized the idea of another distinctively separate realm, yet not so far away. George MacDonald, in his fantasy *Lilith,* had the hero climb through a mirror into the "other" universe.[13] When he wanted to know where he was in comparison to the universe which he was familiar with, it was explained that his favorite chair and the fireplace in his library at home were located about where a bush was standing nearby! Lewis Carroll borrowed MacDonald's idea when he had Alice go "through the looking-glass." In C.S. Lewis' "Narnia" series, the children go into another realm through the back of a wardrobe.[14] Various science fiction writers have enjoyed playing with the concept of "parallel universes" which were accessible by means of a "hole" in the air or a fantastic machine. The point being made was that there is more to reality than *this* universe of space and time and material, and the idea of other very real and tangible realms of existence independent of this universe need not be written off as ridiculous. There is a natural realm — we are in it — and there is a supernatural realm which is entirely separate and distinct from the natural, not just visible and invisible entities in the same realm.

The idea of a "hole" or window leading from this universe into the other one isn't so far-fetched either. That seems to be what happened when Stephen, the first Christian martyr, was being stoned to death. He saw "heaven opened" — *the sky had a hole in it!* — and through that hole he could see Jesus, who was close enough to be identified as to who He was and what He was doing — standing at the right side of God's throne.[15] Just a vision? Do you want to say that Stephen did *not* see heaven opened, because it actually took place in his mind?

We might also have to pass off the experience of the Apostle John on the Island of Patmos when he saw "a door standing open in heaven" and heard a voice saying for him to "come up here."[16] He did it, of course — how do you disobey a command like that? — but was "in the spirit" when he went into the other realm. Did all this only take place in his head?

The Apostle Paul had a similar experience during which he was not at all certain as to whether his mind was in his body or whether he left his body behind during the trip into "the third heaven." How could he have been sure if it seemed to him that he was not a disembodied spirit and yet he was aware that the flesh and blood bodies of this universe are incompatible with the heavenly environment?[17]

One thing is certain: in each case these men were convinced of the reality of their experience. Another significant feature is that their ability to see into or enter that other realm was beyond their normal capabilities. The door had to be opened, as it were, from the other side!

The Bible actually has very little to say about people of this universe having had direct access to the other Heavenly one, but quite commonly there are references to persons — angels from over-there coming here. Very often they looked just like human beings while on other occasions their physical superiority was obvious. They could remain invisible and do all kinds of things which no human being is capable of doing. In certain ways there were similarities between angels and the resurrected Christ. At least, both displayed powers and abilities which were unearthly.[18]

This fascinating fact leads us back to our imaginary viewing-spot in Heaven at the creation of this universe.

Especially interesting to observers would have been the fact that in this new realm, the natural universe is an object — a World — in size about midway between the tiny atoms on one extreme and the monstrous galaxies on the other. Because of the detail and care which God has invested in this World it is evident that He has something extra-special in mind for it. To the limits of our observation it seems to be unique. Eventually it will be the home of a unique being, Man, but before that occurs, an "inter-cosmic" tragedy takes place. We must label it that because both universes are involved.

We're still suffering from the results of that tragedy, which came as a result of the angels, and one in particular who was not content to wait and watch and serve the purposes of God the Father.

"AND THERE WAS WAR..."

Imagine how the angels revelled in the new universe when God first created it! Can't you picture them "diving" out of heaven into the natural universe and then back again? Can you see them cavorting joyfully through space, exploring the galaxies, enjoying the sensation of being supernatural creatures in a realm made up of totally different materials than their own bodies? Imagine them singing out with delight at each new discovery of the beauties and marvels which their God had made.[19] Perhaps they experimented with their abilities, designed for normal functioning in the environment of the supernatural realm, to learn what they were capable of doing in this new environment. They learned that they were, so to speak, "amphibious," able to survive and function in either realm although they were composed of the "stuff" of the supernatural. What was just normal ability in Heaven was super in the natural universe. They liked it. It was good — God designed and constructed it, didn't He? And He made it for the specific purpose of being inhabited.[20]

The reason God created the angels in the first place was at least partially in anticipation of the eventual creation of Man. The angels were to be God's agents for the benefit of His people in this universe.[21] However, an angel is an intelligent, acting and thinking being in its own right, and this is where the problem came in.

There was one particular angel who must have been almost the grandest and greatest of all Heaven's vast, created population. We know his name: it was Lucifer.[22] Apart from His power and grandeur, about the only admirable thing I can think to say about him at this stage is that he was innovative — and in this case even that wasn't good. That which he did which had never been done before was to rebel against God![23] How could he do it? It just seems unthinkable that anyone who, like Lucifer, had experienced the very presence of God could turn against Him. It does demonstrate something about angels though, which is that they are creatures with minds and wills and self-consciousness. God knew that Lucifer would rebel but for reasons of His own He not only allowed it to happen but had already taken that fact into consideration in His ultimate plan. Possibly a third of the population of the heavenly universe joined Lucifer in the civil war which followed.[24]

You think Heaven can't have problems? There was *war* there! If we can't even imagine how wonderful Heaven is, how can we conceive of the magnitude of that war? Of course the final result was a foregone conclusion even if the eventual losers thought they had a fighting chance. Certainly Lucifer and his mob knew they were not idealistically fighting to uphold a just and noble cause!

They were cast out. Crushed by the overwhelming might of General Michael, the commander of the armies of God, Lucifer and every last one of his underlings were driven out of the wonderful supernatural universe in which they had been created. Into exile they were driven, beings composed of the "stuff" of the supernatural universe, aliens to the natural universe.[25] If we could have been there on Heaven's side, watching, we might have witnessed space itself burst open, and a cascading host of conquered angels forced, flung or fleeing, screaming with rage and pain and hatred.

It appears that for a while Lucifer and perhaps the occasional member of his company were temporarily allowed to return to the other side, as and when it suited God's purposes.

There is cause to believe however, that even that temporary access is no longer permitted, and until their final miserable destiny, these creatures experience the very limited freedom of inhabiting a world for which they were not designed.[26] We were not there and God has only seen fit to give us a few clues, but this scenario seems to fit the limited information we have.[27]

God doesn't do anything without reason; thus, in spite of this, the plan and purpose of God carries on. It was God's original intention that the world be the home of man. So God creates man. He makes him of the material of this natural universe, but with certain important qualities which are based on the supernatural universe and on God Himself.[28] Most specifically He made man with a mind and will which would not just be locked into the machine of the natural system. In this way man was able himself to be an agent for bringing about changes in what would otherwise have been a mindless chain of natural events as the various parts and particles of the universe acted and reacted to one another. Man is capable of filling a special function. God can assign him duties to perform. It is man's task to use his mind to figure out how things work in nature and to learn how to control nature. Man, God tells him, is to be God's foreman and superintendent over nature.[29]

God made this universe to be a setting in which He desires certain events to occur. In order for this to happen it was necessary that mindless nature not just be allowed to run its course. Nature provides the setting in which these events take place. The framework is an orderly and potentially predictable system of natural cause-and-effect. Into this blind and otherwise deterministic machine certain *Agents of Change* are capable of bringing about events that are not natural; that is, things occur in this natural universe which would not ever have happened if the impersonal forces of nature were left undisturbed.

The first Agent of Change is, of course, God, who is able if He chooses, to reach in from His supernatural realm and cause something to happen which would not have occurred had nature been allowed to run its course.[30] God does much of His intervening through the agency of His angels which were created for that purpose.

The next agent of change is man. He is like God in this aspect. He can think and choose and act to bring about events, things which would not otherwise happen.[31] He is God's co-agent.

Man hasn't upheld his end of things at all well. So many of the changes which man has been instrumental in bringing about have not been in line with the duties assigned him by God. Like Satan before him, he's attempted to run things his own way. As a result, each individual man, as an agent of change, has to decide whether he'll assume that special function which God has for him, or whether he'll try to go it without God.[32] In any case, man is involved, and whatever he chooses to do will have a bearing on the way things turn out.

And then there is Lucifer and his angels! Since being thrown out his name has been changed. He is no longer known by his heavenly name, but is Satan — the devil. We don't usually call Satan's crowd "angels" now either. They are "demons" or "evil spirits."[33]

Satan also has plans for this world. He wanted a kingdom, like God. If he can't have heaven, he'll take the world! If he can't be god there, he'll be god here. This world, he is determined, will be his realm. God has the supernatural universe but the natural universe will be under totally different domination if Satan has his way.[34]

Let's review what we've uncovered so far. There are two great systems or universes, the supernatural and the natural. In the natural universe we can see that the framework is like a vast cosmic machine, mindlessly and impersonally functioning on the basis of the natural laws of cause-and-effect. It might be compared to a vast clock which is ticking methodically on but, if left to itself, will eventually run down.[35] But it is not left to itself because there are *agents of change* who are causing things to happen which would not otherwise have ever come about. There is the supernatural Agent, God. Then there is the natural agent, man. And then there is Satan and company — supernatural in composition but shut up here with us in this natural universe and working in antagonism against the supernatural realm.

This deserves some special and serious thought. There are two realms, and two legal or authorized agencies of change, one belonging to each realm. But then we see an illegal agency of change, expelled and alienated from its home realm and operating, temporarily, in a realm which is foreign to its own nature. That means that in this world there are things happening which are not natural or right or in harmony with the way God designed nature to function. These events appear to the natural observer to be supernatural. Yet they are not from God nor do these events have their origin in the supernatural realm, but from powers at work entirely within this universe. We have identified and named the Supernatural; we have defined the Natural. Now let's name this other source of power or activity. Let's call it — the *MYSTICAL*.[36]

Notes — Chapter One

1. Psa. 90:2; Prov. 8:22-31
2. Isa. 66:1; Heb. 11:16
3. I Kings 8:30; Matt. 5:16, 6:9

4. Psa. 11:4; Isa. 57:15; Matt. 5:34; John 14:2
5. Isa. 6:2; Matt. 22:30, 26:53; Heb. 12:22; Rev. 5:11
6. Ezek. 1:5ff.; Rev. 4:6-8
7. I Kings 8:27; Psa. 139:7-12
8. Mark 16:9-19; Luke 24:36-51; John 20:14-29, 21:1-14; Acts 10:40,41
9. I Cor. 14:33
10. —Through the Son: John 1:3-10; I Cor. 8:6; Col. 1:16; Heb. 1:2
    —Creation "spoken" into existence: Gen. 1:3,6,9,11,14,20,24; Psa. 33:6, 148:5; Heb. 11:3
11. Psa. 145:1-13; Isa. 40:26; Matt. 26:64
12. Isa. 34:4, 51:6; Psa. 102:25,26; Luke 21:33; II Pet. 3:10. Recognition that the universe is "running down" is built into Newton's "Laws of Thermodynamics," those principles which state that entropy increases in a closed system, a concept never demonstrated erroneous by an example to the contrary.
13. MACDONALD, George: *Lilith*; London: Lion Publishing, 1895
14. LEWIS, C.S.: *The Lion, the Witch and the Wardrobe*; London: Geoffrey Bles, 1950
15. Acts 7:54-60
16. Rev. 4:1,2
17. I Cor. 15:39-50; II Cor. 5:1-8, 12:1-4
18. Num. 22:31; Judges 13:3-21; II Kings 6:17; Luke 1:11
19. Job 38:7; Prov. 8:30,31
20. Gen. 1:26; Psa. 115:16; Isa. 45:18
21. Psa. 91:11,12; Matt. 18: 10; Mark 1:13; Luke 16:22; Heb. 1:14
22. Isa. 14:12-15; Lucifer: Heb. "the shining one." This is a somewhat problematic prophecy, but seems to refer to Satan.
23. Ezek. 28:12-19; II Pet. 2:4; Jude 6
24. Luke 10:18; Rev. 12:3-12
25. This is why they are called "spirits"; they come from the Spiritual realm, are non-substantial and like "breath" to us. Eph. 6:12
26. Job 1:6, 2:1; I Sam. 16:14; I Kings 22:19-23; Luke 22:31; John 12:31; Rev. 12:10
27. Gen. 1:2, a preferred translation would be, " . . . but the earth had become a ruin and a desolation." For a thorough study of this subject see:
    CONSTANCE, Arthur C.: *Without Form and Void*; Brockville, Ont., Canada: Doorway Papers, 1970
    "Gap Theories," or the idea of a "Pre-Adamic Renovation" in any of their various forms, are not essential concepts to the world-view being presented, but can provide useful explanatory models.
28. Gen. 1:27, 2:7; Psa. 8:3-5
29. Gen. 1:28, 2:15; Psa. 8:6
30. E.g. Gen. 6:7,17; Job 23:13; Isa. 45:7; Jonah 1:4,17, 4:6-8

31. Gen. 11:6; Isa. 3:10
32. Gen. 3:1-6; Josh. 24:14,15; Isa. 53:6
33. Job 1:6; Matt. 12:43-45; Luke 8:12; Rev. 12:9, 16:13,14
34. Dan. 10:12,13; Matt. 13:18,19, 25:41; II Cor. 4:4; Eph. 2:2, 6:11,12; II Thess. 2:7-10; I Pet. 5:8; Rev. 2:13
35. The image of the universe as "clockwork," as predictable in principle as a machine, has lost its popularity. Not only atoms but the various subatomic particles which have been discovered give the appearance of randomness. However this irregularity is in principle no different than the fact that, on a macro level, each galaxy differs from every other, yet all function uniformly on the same natural principles. The point is, that the universe is a mindless system and works with a consistency which comes from the fact that all of its parts are equally subject to the same finite forces and are compounds of the same finite substances. At least the contrary has not been established. The so-called "uncertainty principle" will be discussed later.
36. Matt. 24:24; Acts 8:6,7, 16:16-19; II Thess. 2:9; Rev. 13:13,14, 16:14, 18:23, 19:20

2

"So What?"

An interesting story, you might say, and basically true *if* you happen to accept the Biblical account as reliable. But what difference does it make? How can this have any practical bearing on our day-by-day existence?

WHY "WORLD-VIEWS" MATTER

Actually it makes all the difference in the world. This is a "World-view," this "story" which the previous chapter has given. It's a partial account of a *belief-system*, one particular set of ideas about what exists, what doesn't exist and what causes things to happen as they do. If you should agree with these ideas even in their general form, this in turn will strongly influence the way you explain your experiences in life, and all the events and objects which you observe around you.

THE SUPERNATURALIST

If it should be that the account in the previous chapters fits the facts as you see them, your interpretation of everything will follow a certain pattern: You will believe in the natural world as a place where the general course of events can be credited to the impersonal and consistent functioning of nature. You will also believe that man is a unique being with the ability to truly think and act and do things which change the course of history. You will also believe that there are some mystical powers which really exist and are not in every case just superstition — which is not to say that there aren't plenty of superstitions! — and that these mystical powers are not good and are to be avoided and treated as enemies.[1] And then you will believe that this natural system is open to influence from the outside by God and His supernatural universe. In addition to believing all these things, if you are a Biblically-oriented Christian, you will place number-one importance on this Supernatural aspect, because God is Number One, and your world-view could be labeled as *supernaturalistic.*[2]

If you reject this previous world-view in favor of another belief-system, your interpretations of reality will naturally be very different as well.

NATURALISM

For example, if you hold to a *naturalistic* world-view, this means that you reject the influence of both the supernatural and the mystical as being "agents of change" in nature, and probably you'll reject their existence entirely. You might call yourself a materialist, an atheist, an agnostic or a deist. You will credit everything that happens as being the exclusive result of natural laws. You would say, "In each and every case *without exception, if* it were possible to have enough information, we could understand that this event has an entirely natural explanation."³

MYSTICISM

If you hold to a *Mystical* world-view, you might believe that the universe is permeated with a vital and all-pervasive force which controls everything that happens. This force might be pictured as a cosmic mind or consciousness, or as one or many spirit beings. You might believe that "God" and the "laws of nature" are one and the same. As likely as not, if you lean toward a mystical way of thinking, you won't have spent much time analyzing what you believe, or putting your ideas into words. Rather than explaining, you'll just "know" what you believe.

The mystical way of looking at things is usually about the same as if we were to mix the supernatural realm and the natural realm into one: to change our understanding of God from seeing Him as a special Person, Some**one**, into an impersonal force, Some**thing**. Then we would believe that only one universe, not two, exists, but that this one universe has an immaterial side as well as a material side. Animism, vitalism and pantheism are terms which you will hear in this context. The mystical view, in various forms, is most commonly associated with the Eastern religions, spiritualism and occultism, but this is only the "tip of the iceberg."

To simplify — or oversimplify — for the sake of illustration, the following classifications would usually be accurate and appropriate:

— Supernaturalism tends to lean toward Christianity
— Naturalism tends to lean toward atheism
— Mysticism tends to lean toward occultism

These may be very broad generalizations, but then generalizations are *generally* true! The *tendency* in each case is to drift in that particular direction.

UNSYSTEMATIC BELIEF-SYSTEMS

Of course, the average person doesn't know where he fits in for sure, and that might well be because his world-view is somewhat of a mixture. On one occasion, he thinks like a naturalist, rejecting anything which can't be explained scientifically as superstition. However, he will go to church, argue that God exists and claim to believe in prayer. And to top it off, he just might be found to believe in Extrasensory Perception or that his mother actually did see the ghost of Grandpa standing at the foot of her hospital bed. Inconsistency is almost certain to be the result when a person doesn't know for sure what he does believe or why.[4] "Belief-systems" on the average, seem to be very unsystematic. Could you possibly be an "existentialist" without even knowing what an existentialist is? Could it actually be mystical powers which you believe in, while thinking you are believing in the supernatural? Calling macaroni casserole a steak doesn't change what it is!

To find a real, true, honest-to-goodness atheist is not all that easy. Even those who claim to be totally naturalistic in their beliefs are usually inconsistent here and there, especially in times of stress. It also appears to be easier to be an atheist while one is living than while dying.

There are probably more agnostics — people who don't know whether or not there is a supernatural realm, and think it's impossible to know — than there are actual atheists.

Mystics on the other hand are not about to become an endangered species! It's been estimated that possibly as much as eighty-five percent of the present world population is animistic, in a somewhat broad definition of the term. Animism is one of the sub-divisions of mysticism as we're using the word.[5]

THE SLIDE TOWARD MYSTICISM

Of great significance is the fact that both the naturalistic and Christian segments of society are sliding into mysticism, which is the primary reason for the existence of this book.[6] A movement towards the mystical is taking place in naturalism because it looks like the only way out of a philosophical dilemma in which they've found themselves — other than by becoming supernaturalists. They usually prefer to avoid using the word "mystical" however and to go for something more objective and scientific-sounding, such as "paranormal."

Christians, in considerable numbers, are moving into mysticism without appearing to notice the transition, in part because they don't know the difference!

The statements which have just been made are apt to provoke strong and negative reactions on two main counts. First, some people will object that no significant "slide" toward mysticism is taking place. Secondly, others will contend that the upsurge in "spiritual" awareness is a good thing, the result of a world-wide "revival." The issue is too important to slough off, so we'll have to look more deeply into the trends which are taking place and analyze what the implications are to us. What really is happening, why is it happening, and what does it all mean?

Notes — Chapter Two

1. Lev. 20:6,27; Deut. 18:9-15; Isa. 8:19-22; Gal. 5:19-21; Eph. 6:10-19; James 4:7; I Pet. 5:8,9; Rev. 21:8
2. Matt. 16:23; Phil. 3:14; Col. 3:1-4; I Tim. 4:10; Heb. 12:2
3. II Pet. 3:3-7
4. I Cor. 10:19-21 — an example of having one foot in the supernatural-

istic camp and one in the mystical.

5. "Animism" is a belief in spirits, both the spirits of dead people as well as others which have no human origin. Page 5,

NIDA, Eugene and SMALLEY, W.A.: *Introducing Animism*; NY: Fellowship Press, 1959

6. The books of Francis A. Schaeffer have warned against this phenomenon since 1968, and before that, C.S. Lewis hinted at his concern in such books as *That Hideous Strength*. About 1952, a professor friend of this author indicated that he expected liberal theology to go mystical because, he said, "that's the only direction they can go!"

3

"And He Went Forth, Not Knowing Whither He Was Going..."

THE DARK-AGE REVIVAL

Trends usually take place without the majority of people even being aware that anything of importance is transpiring. In this particular case I wish it were possible to simply sit back and objectively enjoy world-wide cultural change in progress. But it isn't progress, it's *regress*, a return to a mind-set which held control of men's thinking, stifling both spiritual and material progress long centuries ago.

To best understand today and tomorrow, we need to understand something about yesterday. Although the trend is worldwide, for our purposes it will be sufficient if we look back to medieval Europe. Suppose we were able to join a tour group for a trip around the Continent as it was eight hundred years ago. Our major impressions of the Europe of that era could probably be summed up in the words "primitive" and "superstitious." The differences which we would have seen would go much deeper than the terrible poverty of the average person. It would almost certainly be the attitudes and ways of thinking of the people which would be most foreign and shocking to us. Primitive and superstitious, we would conclude, are fully appropriate labels.

Suppose for example that an event were to occur, as would often be the case, which could not be explained by the pre-technical and pre-scientific knowledge of an individual. Any such experience would immediately be attributed to God — or the devil or spirits or witchcraft or some other invisible power. When danger threatened, one might instinctively make a religious sign with one hand, reach for a good luck charm or rosary with the other and, while invoking the name of God or patron saint, run to hang a sprig of garlic in the window. Any event of significance or potential significance would immediately be credited to some kind of mystical power or personality.

Their cultural world-view simply didn't have a place for presuming that we live in a natural system with a basic framework of impersonal and mechanistic cause-and-effect. No clear-cut distinction would have been made between natural and supernatural agencies of causation.

A man back in those days didn't see himself as an agent of change, an innovator. Mostly, he didn't want to rock the boat and upset God or any other of the invisible powers who held the exclusive power or right to determine the future of men and nations. Kings or religious leaders, or other special people might be allowed to share in such decisions, but not the ordinary man.[1]

No one would have questioned the existence of miracles and visions, but these were not perceived as being supernatural intervention into nature's normal system. Rather they were the evidence that some people had within them special abilities, psychical powers, either for the good or evil. Heaven itself wasn't viewed so much as being an entirely separate realm as it was some place "up there," rather like the ancient Greeks surmised, from which perfected spirits watched and sometimes intervened. The universe was full of mystery.

Even those great churches and magnificent works of art which were to be found, would only make these facts more glaring and conspicuous. Have you ever gazed in awe at the magnificence of a thousand-year-old cathedral? Did you perhaps notice gargoyles, hideous stone faces, protruding from high up on the walls? These objects were certainly not for the sake of beauty! They were to frighten away the evil spirits. The art was also more than merely decorative. If not directly symbolic and mystical, the object was the creation of an atmosphere, perhaps of grandeur and power, or of mystery and awe.

Don't think the case is being ridiculously overstated. Most twentieth-century people would be hard-pressed to realize that their ancestors would have considered the concept of natural cause-and-effect to be heresy — if they had been capable of considering it at all. While one reader will be criticizing this as "overstatement," there will almost certainly be another who will be perfectly comfortable with the medieval belief that every object and eventuality is part of — or should be part of — a divine design or mystical manipulation of which they are but a component.

The point of the matter is that a few hundred years ago in Europe we would have found nearly perfect consensus of thought without this issue even being opened for discussion. Because of the belief-system and thinking styles produced by their religion — an ornate structure of Christianity and tradition built over remnants of ancient paganism — it simply was the way the world worked, as far as they were concerned. Their interpretation of reality — their world-view — was deeply mystical.

THE BELIEF IN PROGRESS

However, one really important advantage the medieval European had going for him, also the outgrowth of Christianity, was the demand for repentance. This encouraged within him a spirit of self-criticism and the assumption that his life could and ought to be better than it was — and a belief in progress. Nowhere else in the world was such a way of thinking to be found.[2]

In the course of time this attitude became a major contributing factor in a clash with the religious Establishment — the very system which had been encouraging him to repent and be dissatisfied with himself, not with the Establishment. Every establishment wants to preserve the *status quo* and to resist those individuals who want change and improvement and want it now! As it turned out, the Immovable Object had helped to create an Irresistible Force.

Notes — Chapter Three

1. Mystical societies tend to be "non-differentiated." That is, simplistically, not divided up into specialties on the basis of technical or administrative skills, religion, etc. Power in every area of life is in the hands of a hierarchy which is believed to influence or be influenced by the spirit world. Government and religion work hand-in-hand. The "divine right of kings" is an example.
2. It's significant that Darwinism and Hegelian Idealism, both assuming inevitable progression, have come from the Western mind-set. See also:

    ROBERTS, J.M.: *The Triumph of the West*; London: British Broadcasting Corporation, 1985

4

"Let Them Have Dominion . . ."

REALITY RE-DISCOVERED

Self-criticism naturally leads to criticism of society in general. Various courageous men began to ask embarrassing questions and then to challenge and even deny some of the basic assumptions of the medieval society's world-view. Of course it was not really new since it had been in there all the time, forgotten and waiting to be re-discovered. This was the supernaturalistic world-view which was laid out pictorially in the earlier parts of this book.

The upshot as far as the medieval church was concerned was the Protestant Reformation, but the changes which ensued were so far-reaching and all-pervasive that the whole world was eventually to be altered in very visible and material ways. The changes in thinking were not only *away* from a mystical world-view, but were *toward* a clear delineation between the supernatural universe and the natural universe.

MAKING NATURE NATURAL

Let's remember that the mystical way of thinking is like mixing the two realms together into one. To mystical thinkers, supernatural entities both good and bad are seen as so completely permeating everyday life, that they are believed to be the most important key to explaining any significant event. Causation in any particular case is automatically assumed to be based on invisible super powers, or persons, which are believed to be constantly surrounding us.

A very important element in restoring an accurate and Biblical world-view was the understanding that a clear-cut distinction existed between supernatural places, personages and powers, and the natural system, with mystical powers explaining relatively little. The framework of man's existence was once again recognized as that of a natural, normal universe which carries itself on in an orderly and predictable way, because God made it that way. By returning to the Biblical world-view, they had succeeded in dividing the mixed-up mystical universe into separate natural and supernatural realms.

GETTING BACK ON THE JOB AGAIN

Another fact was re-discovered by these pioneer thinkers of our present era. Man had a God-given task to perform. He was intended to be an agent of change. He was to learn to understand, control, take care of, and utilize the natural universe for the ultimate glory of God. To intervene in the course of events was now altogether reasonable in light of this newly-recovered understanding. The natural universe was exactly that — a natural system. To become active agents of change in the natural order was clearly not blasphemous, heretical or immoral, since they had a divine mandate and the only mysterious and invisible powers which might be disturbed by it were the mystical "illegal trespassers," anyway.

When people who believed this way began to investigate and examine nature with curious and prying eyes, they fully expected to find that the system was orderly, consistent and in general, understandable. After all, had it not been created by the God of order and were not they themselves rational agents of change?

"Therefore," they said, "let's start taking things apart to find out what makes them tick!" In a very real way, the restoration of the Biblical world-view and the style of thinking which went with it opened the door to this age of science and technology.[1]

KEEPING GOD IN THE PICTURE

It must never be thought that the men who re-discovered that nature was natural were ignoring God. What they did occurred *because* they so truly believed in Him. The natural universe was still open for God to intervene wherever and whenever He chose.

A little farther along the line we'll discuss the important issue of whether or not God has a specific plan for each person's life and to what extent people have a right to set goals and try to influence future events. The point we are making here is that while — and because — the people who started off the modern world on its current path were Christians, they saw that the basic framework within which normal activities take place is purely natural. When they looked at each event or experience they first presumed natural causation and with each new discovery this presumption was reaffirmed. Nature, they found, was very natural indeed.[2]

To sum up this point, we have discovered two of the major results of returning to the Biblical world-view:

—The belief in the naturalistic and normal functioning of nature as a system;

—The obligation which man has to work hard at controlling and using nature to God's glory.

Wherever Bible-believing Christians have gone throughout the world, they have carried these assumptions with them, often not aware that they were engaging people who held vastly different beliefs, values and attitudes.

These two assumptions — and of course the world-view from which they flow — have obviously been the strongest in those countries most influenced by the Protestant Reformation. Here's a mental exercise to demonstrate this fact: In your mind, rank the various countries of the world in order, on the basis of standard of living, technological development, human rights and educational and health standards. Then label them on the basis of predominate religious influence. See what I mean?

Something else we have to admit is that present-day science, education, technology and the like, while not especially noted for being the strongest supporters of the Christian faith are in fact the products of a biblical faith. The way this present-day antagonism came about will be seen next as we continue our brief analysis of history.

Notes — Chapter Four

1. The modern scientific world, with all of its concomitants, arose out of the Christian milieu; more specifically, it was the Protestant Reformation, in restoring a Biblical cosmology, which made possible the thinking from which the modern international power structure developed. This has been testified to by an imposing list of scholars, both Christian and non-Christian, including Max Weber, Alfred North Whitehead, Harvey Cox, Talcott Parsons, J.R.

Oppenheimer, J. Gresham Machen, Carl F.H. Henry and Francis A. Schaeffer.

2. The rather shocking statement that Christianity is the only avowedly materialistic and naturalistic religion is entirely understandable and obviously true when compared with all mystical thinking. Material and nature have real meaning only when understood in the light of the immaterial and supernatural. See:

OWEN, D.R.G.: *Body and Soul*; Philadelphia: Westminster Press, 1956

5

"They Did Not Like To Retain God In Their Knowledge..."

"Naturalism" as as world-view is, simply, the belief that nature is everything. There is no separate realm of the supernatural. Neither are there any invisible mystical powers. Mass in motion through space in time — that sums up the totality of existence in the mind of the genuine naturalist. There once were quite a few of them around.[1]

THE RISE OF NATURALISM

Where did naturalism come from? It seems to be almost entirely a modern innovation, something which would almost appear to have just *happened* to science and philosophy when they become too preoccupied with the way the natural system functions. Having conceptually separated everything that exists into the two distinct realms of a supernatural universe and a natural universe, a chain reaction of ideas took place.

Step One was the separation of nature into its own system which is self-perpetuating — at least for a limited period of time — and which appears to function by mechanistic-type laws.

Step Two was the recognition that man himself is clearly a natural being who lives and functions entirely within the natural realm. He cannot see, feel, taste, touch or smell the supernatural realm. Man cannot manipulate the supernatural, only the natural. That's all he has power to get at. Man's own research, it was concluded, could never reveal anything except the characteristics of the natural universe around him.[2]

Step Three was the discovery of just how truly systematic nature actually is. The assumption of natural causation works when put into practice. The development of increasingly more complex and sophisticated technology is the evidence. The more we know, the more accurately we are able to predict and sometimes control what happens next.

Suppose for example that a man has a thousand questions about why certain things happen as they do. And suppose that in each of nine-hundred and ninety of these cases he discovers that the explanation is purely the functioning of natural principles. What conclusion is he likely to draw regarding the remaing ten unsolved questions? He will almost cerytainly assume that when he gets enough facts, these also will be found to have rational, natural explanations.

Step Four was to accumulate more and more knowledge about nature, and to develop better tools and instruments for the study and manipulation of man's environment. The more men learned, the more they found that there was to be learned. For every question a scientist answered, another dozen mysteries were exposed. There's just so much waiting to be discovered.

Step Five was to divide the growing pile of information into specialties of study. To study a system, one ought to be systematic. It was discovered that the principles which apply to certain parts of nature do not equally apply to every other part. For example, the principles which apply to living things are not exactly the same as those which apply to non-living things. The obvious solution was to select one distinct part of nature as a field of study, since no one can know everything about everything. A specialist, it's been said, is someone who knows more and more about less and less. Thus some men become botanists while others become zoologists. Others were astronomers or chemists or geologists or meteorologists. Some became physicists and others psychologists.

There is a problem however with being a specialist. It's such a temptation to get lost in the details and forget where you came from. In this case, the result of such total absorption in the intricacies of each of the different disciplines, each of the many subdivisions of study, was to ignore the big picture. It was no longer the study of nature in the light of its supernatural origin. It was just nature for nature's sake.

Step Six occurred when men began putting into words what they had already been practicing. First they tended toward *Deism*, the idea that God exists — theoretically at least — but never ever interferes in the course of natural and human events. There must be a supernatural realm to account for where everything came from originally, but the door between the two realms is now shut tight. We live in a *closed system* in which everything without exception must be explainable on the basis of naturalistic principles.

Deism was pretty much a passing fancy, a transitional stepping-stone on the path to full-fledged naturalism. Man just didn't see that he had any more need for God, and to be honest didn't really want to have to take God into consideration. Just as Satan did originally, man wants to be autonomous and center everything around himself. So, he moved on from deism to full-blown naturalism. Nature, he declared, is the sum of all things.

NATURALISM: ACCIDENTALLY ON PURPOSE?

Do these six steps sufficiently account for the transition from a Biblical world-view to a naturalistic one? Was it all just an unconscious drift, the result of preoccupation with the search for more knowledge and for the principles by which nature functions? Perhaps this was the case for the average person who was involved in education, engineering or whatever. Most people tend to accept what they are taught although with a touch of skepticism, then chuck out most of the philosophy and get down to work. Getting the job done on a practical level is all most folks are concerned about. If there are a few inconsistencies in their belief-systems, it's no big deal. Most of us can live comfortably with a certain amount of ambiguity.

There were others at work however who were not in the least unaware of the trend, but who were deliberately and consciously attempting to bring about the exclusion of everything supernatural, especially a personal God. Their main method of approach was to attack man himself.

Man, they insisted, is simply an animal, no more. They challenged the supernaturalist's belief that man is unique in nature, the one natural agent of change in an otherwise mechanistic and impersonal universe. Man, they said, simply has a more highly evolved brain than any other animal. Isn't it evident that we ought to explain the *origin* of man, the earth and the entire universe on the basis of the same principles by which we are attempting to explain the *present functioning* of nature, that is, on the exclusive basis of natural cause-and-effect? "Exclusive" means excluding God and the supernatural from the explanation. Surely if the assumption of naturalism is true now, it has always been true, and if this is the case, how can we believe that the earth was deliberately created? To be consistent must we not assume that nature comes from nature, that everything just happened — evolved — as a result of a vast amount of time, the impersonal functioning of nature and pure chance?

NATURALISM: A CHRISTIAN HERESY

Yet there remained certain important ways of thinking which naturalism had inherited from its Christian parentage. The very name for the philosophy, "Naturalism," would be nonsense unless one understands the idea — the possibility perhaps? — of the *supernatural!* The thinking of the naturalist still begins by making a clear distinction between the concept of the natural and the supernatural. He doesn't say, "God? What does that word mean?" He knows perfectly well what he means if he argues against the existence of anything other than the closed system of nature. Secondly, he still optimistically believes that progress for mankind is possible if not inevitable.

Notes — Chapter Five

1. For a Christian view on anti-supernaturalistic naturalism, see:
LEWIS, C.S.: *Miracles: A Preliminary Study*; NY: MacMillan Co. 1947
SCHAEFFER, Francis A.: *The God Who Is There*; Chicago: Inter-Varsity Press, 1968
2. Various philosophers, in the process of excluding the supernatural from their world-view, concluded that all human knowledge must come through the senses, and the senses only register things from nature. Therefore, it is impossible to have any knowledge of anything supernatural.

6

How Naturalism Became Unnatural

THE EVOLUTION OF NATURALISM . . .

In Chapter Three it was stated that the changing trends in thinking which are taking place are in actuality, a regression, a return to a world-view which is many centuries old. Naturalism, with its corrupted version of the Biblical world-view, may eventually suffer the same fate as the dinosaur. It must adapt or die, but if it adapts it will no longer be naturalism. Naturalism is going mystical.

This is indeed a pity, not because the demise of classical naturalism is any great loss, but because its replacement is inferior, terribly destructive and far more difficult to deal with. At least the Christian and the naturalist could argue. The principles of both the Christian faith and the naturalistic faith are communicable. They spoke in terms the other could understand even while disagreeing, and they reasoned in the same objective and analytical way. Mysticism differs in that it ultimately becomes immune to debate, discussion, or even normal human communication. Its nature is to be mysterious and not really explainable. It must, it is claimed, be personally experienced and practiced to be fully understood.

"I OBJECT — I THINK!"

"Hold on now!" you say. "That's no different than being a 'born-again' Christian, is it? How can we condemn that?" Don't throw the book away yet! Right now it's naturalism and mysticism that we're comparing. How this has, in turn, influenced Christianity will come in later. Besides, we need to remember a lesson from history. The medieval church, because it was so mystical, also made the mistake of charging those who restored the Biblical world-view with heresy, so let's withhold final condemnation until we've followed the course of events a little further.

THE SLOW DEATH OF NATURALISTIC PHILOSOPHY

The question at hand is, how did naturalism begin its slide into mysticism?

Actually, the seeds of naturalism's own destruction were there from the beginning, built right in. At first it had all looked so neat and tidy: natural causation was sufficient to explain everything. God and the supernatural were irrelevant and outmoded. Then the flaws in the structure began to show up, until eventually, only the most stubborn and blind could deny the existence of major problems.

THE BUILT-IN WEAKNESS: THE CLOSED SYSTEM

The origin of these problems lay in the idea that nature is a completely *closed system*. "Closed" was understood to mean that this natural universe is all there is. Everything can and must be explained by natural processes alone. Whatever happens is the product of previous influences, conditions or events within the natural universe. Nature is closed to any outside intervention.

DETERMINISM: NOTHING BUT A MACHINE?

Step One came with the realization that a *truly* closed system seemed to infer *Determinism*. Every event is determined and fixed by those which preceded it. If it were possible to have enough information, everything could be explained or predicted. Nothing just happens by itself. Nothing is independent of the system. The billiard balls are whizzing around on the table, bouncing and whirling and colliding with one another, but every last movement is attributable to the combination of forces which was applied to it. In theory, it could even be explained mathematically if it were possible to get sufficient measurements, which, of course, it isn't. Another way of saying that everything is determined, is to say that nothing in nature is free.[1]

MAN: JUST A COG IN THE MACHINE

Taking *Step Two* was just a matter of being consistent in the application of the principle of determinism. If man is a product of nature, he is also part of the machine, the result of evolutionary history. The impersonal and mindless outworking of nature explains his existence. Nothing in nature is free and man is part of nature. Therefore man is not free.

MAN LOSES HIS MIND!

Step Three looked inescapable: Mankind is locked into

nature mentally as well as physically. His "mind" is no more than the functioning of his brain, that complex bunch of nerve cells inside his head. This came about as a result of accumulated accidents within the evolutionary process. Every thought, every idea, every conclusion or act of will has been predetermined by the multitude of circumstances, (hereditary and environmental), which went before. The committed naturalist doesn't dare allow anything to happen independent of the system, or the door was opened to supernaturalism.[2] Everything must be explainable *holistically* — on the basis of *one* whole — and supernaturalism insists on believing in *two* wholes.

The necessary implications of determinism started out looking rather attractive. If man was not free from nature, determinism at least allowed him to be free of God.

"Ethics" are therefore just an illusion. The word is meaningless. Man is a part of nature and "doing what comes naturally" is the only kind of "right" there is. Man was free to devise his own moral standards, should he so choose, rather than have a God impose His. Man liked the idea of stepping into God's shoes and becoming the final definer of what is "good" and "evil." He could allow his conscience, if any, to be his guide, but didn't have to if he didn't want to.

"Purpose" is another illusion. Man was not deliberately created for a reason, and therefore has no divine obligations. Being only the accidental product of evolution, he, like everything else in nature, just *is*! His only responsibilities and duties are those of his own making and are arbitrary and ultimately insignificant.

One of the greatest illusions of all is the belief that man has a "free-will"! *Nothing* is free, and therefore man only imagines that he is able to make real choices in which he could actually choose to do something else. Of course we all have to *act as if* we have free will, but if we only knew enough we could predict every flutter of an eyelash or flicker of an idea! The comforting thing about this is that man could at last abdicate all responsibility for supposed misbehavior. No free will means no accountability. How convenient! You can't blame me for anything. I'm entirely the product of circumstances over which I have no control!

But if the principle of naturalistic determinism is true, there is one massive illusion which overshadows all the others: *the illusion that man is capable of rational thought.* How can we get the rational from the irrational by accident?

Even the words "rational" and "irrational" become meaningless. If nothing is independent of natural causation, then every impulse of the brain just happened to turn out that way, just as the grains of sand on the seashore just happened to wash up in the way that they did. No reason, no choice, no sense? Nonsense!

NATURALISM CAN'T LIVE WITH ITSELF

Something was obviously wrong with the philosophy. What seemed to be a necessary implication of the theory was obviously not true. No one can actually live as if there is no such thing as rationality without being institutionalized. The concept that all reason and logic are illusory can be played around with as a theory, or might make a good sensationalistic gimmick to gain some notoriety as a professor or artist. The idea won't work in real life, however. The only way one may even attempt to live by it is to have other more realistic and practical people around to take care of us and to keep the wheels of society turning.

And speaking of *inconsistency!* What is more ludicrous than, for example, an "artist" who has smeared, splashed and dribbled paint over a canvas in keeping with his philosophy that life has no meaning or purpose or value, and then hear him pontificate to an enraptured audience on all the *reasons* behind his style!

If the philosophy of naturalism were true, it appeared inescapable that man would not even know it, because knowing is a function of reason. A philosophy of any kind is much more than mere animal awareness of a thing; it is an organized system of ideas about everything. Here we have a concept which leads us to the logical conclusion that concepts and logic don't exist![3]

But concepts and logic *do* exist. Everyone apart from the lunatic fringe knows this is so. What was the naturalist to do? How could he account for the fact that his philosophy makes it awkward — to put it kindly — to even explain how he can think real, meaningful thoughts? How could the naturalist accommodate the fact of human rationality into his system? There had to be a solution.

NATURALISM LOOKS FOR AN ACCEPTABLE ALTERNATIVE

One obvious alternative was to scrap naturalism's closed system altogether and acknowledge the Supernatural, Rational Creator. This makes sense, but since the majority of people don't want to acknowledge God, various other solutions, ingenious and otherwise, have been devised.[4]

Possibly the most popular is to simply ignore the problem. The cheap solution, if one can't justify his position, is to simply declare it to be so and scoff at anyone who would challenge it: "The universe is a totally naturalist closed system; reason and logic do exist and that's a fact. Don't be ridiculous by making it into a problem. We just don't have enough information yet and perhaps never will, but facts are facts. I'm a practical man so just let me get on with my work."

It was during my final exchange with a Soviet scientist-philosopher at the end of a long evening's discussion that a classic example of this form of ignorance occurred:

"Well," I concluded, "It all really comes down to one thing. If you were right you wouldn't even know it!"

He blandly smiled at me and shrugged his shoulders. "That's one of those questions we hope to have an answer to some day."

Another idea which has been making the rounds is known as the "Uncertainty Principle," the conclusion that ultimately, at least on the sub-atomic level, there is no mechanistic cause-and-effect process in nature, but instead an unpredictable randomness exists. Perhaps this helps explain how man can have a mind that isn't locked into a deterministic system. But it doesn't help very much because nothing has really changed. Natural causation is still there but it is just too small and complex to be discernable. And, man is still seen as an accidental product of impersonal and irrational forces in nature. I suppose some people get a little comfort from playing around with the possibilities which the uncertainty principle offers in the hope that it might provide a way out of the dilemma — which it doesn't. There isn't much real comfort to be found in uncertainty!

Maybe all of this doesn't seem like a problem to you. It's no problem for most folks. No! This last statement isn't true. It *is* an actual and significant problem, but the majority of people neither know nor care about its existence. You have to be very sure of your world-view and to have thought through the necessary implications of naturalism before you'll even know that the problem is there. Strangely, very few people seem to have actually done this.

It matters plenty to those who want their thinking to agree with reality. And it matters to those who desperately hope that reality will agree with their thinking. Ideas like these are important enough to get emotional about!

The anti-supernaturalists have suggested many different theories to account for man being what he obviously knows that he is, but most of them are unconvincing to say the least. For example, might it not be, says one line of thinking, that life on earth first arrived as garbage thrown out by space travellers in the vastly far-distant past? Did the human race possibly evolve into intelligent beings because of the assistance and manipulation of higher beings from outer space? This is the kind of idea which is propagated by people who consider "reality a crutch for those who can't face science fiction." Doesn't it ever dawn on them that these are no solutions at all? Suppose this truly were the way the human race originated. How did those super-intelligent space creatures evolve *their* minds? All it does is push the problem out into space and farther back in time, but *the problem is still unanswered* just as is the case with the "uncertainty principle." How do we get the rational from the irrational by accident? Ideas like these have some genuine value however: economic value! They sell a fortune in books to the gullible and generate lots of money and notoriety for a few clever authors.

In the final analysis, virtually all of these attempted solutions make the same two assumptions: Everything *is* the result of gradual evolution rather than having been specifically created, and man *does* have real and significant thoughts and feelings. Therefore, the conclusion is, it *is* possible for rationality to come from irrationality because it has done so! *How* it happened hasn't been fully discovered yet, and doesn't really matter to most people.

HOW NATURALISM BECAME SCHIZOPHRENIC

When we start putting together some of the conclusions and observations which we've accumulated, the very interesting results indicate quite clearly how naturalists could look seriously at mysticism as a viable alternative to solve its problem. Carefully and methodically work your way through this chain of ideas and see where it will lead:

- Nature is a closed system.
- Man is part of nature.
- Man is rational.
- Rationality is thus part of the closed system of nature.
- Man's rationality is not free but is limited to nature.
- Man cannot rationally know supernatural things.
- Supernaturalism is not rational.
- Nature provides no basis for purpose, meaning or morality.
- Rationality provides no basis for purpose, meaning or morality.
- Purpose, meaning and morality are not rational.
- Man wants and needs purpose, meaning and moral standards.
- Man believes in the fact of purpose, meaning and morality.
- This belief is not rational.
- Faith is either irrational or beyond the rational.
- Faith is separate and distinct from reasoning.
- Faith belongs to intuition, experience, emotions, feelings.
- Faith belongs to religion.
- Reason belongs to nature.
- Religion and nature are separate and distinct.
- Religious faith and naturalism are not contradictory.
- They are too different to be compared.

TWO WAYS OF "KNOWING"

Where does this leave us at this point? It leaves us leaping back and forth from one philosophical foot to the other, between intellect and intuition, head and heart, facts and feelings, materialism and morality, science and spirit. Not having a supernatural realm to which they could look to give meaning and purpose to all the observations and details around them, naturalists must substitute a non-rational "leap of faith" and then hop from one side to the other as the situation requires.

Now there are two ways of "knowing," a naturalistic way and an intuitive or mystical way; the objective and the subjective, a look outside and a look inside.

LOOKING FOR AN ANSWER WITHIN

The naturalist who has gone through these processes has now changed his philosophy. It isn't pure naturalism any longer but is *Naturalism Plus* — naturalism plus a strong dose of mysticism.

Why "mysticism"? Earlier we gave that name to the Satanic! Obviously we're now using the name for a way of thinking and believing rather than for evil spirits. Just as the mystical beings attempt to substitute themselves for the supernatural, so the mystical way of thinking is a substitute for supernaturalism. The reasoning will become clearer as we continue.

The changed philosophy which is no longer naturalism but not yet mysticism, which we have called "Naturalism Plus," has its own formal title. The name of this philosophy is *Existentialism*. On the one hand — sorry, foot — the existentialist will say, "I know because . . . " and produce his evidences or reasons. When he uses logic he is standing on his naturalistic foot. On the other foot — the mystical one — he will say, "I know because I know because I know!" Two ways of knowing. Two ways of thinking.

We have been using the term *mysticism* very loosely. Trends which are going in that direction are "mystical" to the extent that they have mystical characteristics. We need to remember that actually existentialism is a step on the way into full mysticism. To be more accurate let's label it with the more general term of *psychical* thinking at this stage.[5] The psychical in turn can be subdivided into two categories: the *intuitive* and the *mystical*. The difference between the two will be explained in the next chapter.

Existentialism is a way of thinking which relies on intuition and internal experiences to give value and meaning in life. However there always seems to be a tendency for the psychical-intuitive to drift into the psychical-mystical as time goes on rather than the other way around.

When this step — having two ways of "knowing" — has been taken, it results in man being divided into two parts and divides the universe into two parts. No longer are there two distinct realms, the Supernatural and Natural, with their respective systems, but one universe is divided into two kinds of things. The one kind of thing is subject to the forces of nature. The other kind of thing is forceful but in a totally different way. These are the things which are extensions of the mind and emotions, beyond the senses, deeper than reasoning, more gut-level than brain-level.

This way of thinking is also very much a characteristic of mysticism, except that in more fully-developed mysticism the non-rational side is less vague and vacuous. One of the most noteworthy features of the mystical Eastern religions and other organized forms of mysticism has always been their claim to possess *irrational knowledge* derived from sources other than natural.

If you've been keeping up so far and getting the picture of this changing world-view, perhaps you've already noticed something significant: the whole issue is MAN. How does man account for himself? Does man have a mind that thinks real thoughts? How can we agree on a standard of values which is just right for man? The problems are man-centered all the way. His naturalistic philosophy tells him he is nothing special, just another component of nature, while his instinct tells him he is very special. He dreams up all sorts of ideas to explain and justify his own "mannishness."[6] This explains why we can also label each of these schemes, whatever slant or form it may take, as some variety of *Humanism*. Existentialism is one form of humanism. So is Marxism. So is mysticism, but we'll get into this more later. Where are we so far in this historical odyssey?

We have seen that Naturalism could not live with itself. Some explanation had to be found to give real value and significance to man and his rationality. Man needs purpose and meaning, a reason for his existence. But although many theories had been devised, none would really hold water. The explanations left unsatisfied the deep longings for value and direction which most of us admit that we have. So more and more the trend has come to be to look elsewhere for the answers. Less attempt was — and is — being made to explain, more attempt is being made to "experience." The search has turned inward.

Once we've become aware of the change of direction in the thinking of society as a whole, examples are all around us. Almost inevitably, it will be discovered that our own attitudes will also have been affected in the process. It's very difficult not to be influenced by something so all-pervasive. This shift of emphasis has spread throughout education, the media, community and social services, religion, you name it. The attitudes and presumptions are to be seen everywhere. It would be naive to contend that we have not personally been influenced as well.

AND SUDDENLY THE WHOLE WORLD LOOKS DIFFERENT!

Most of society, of course, has not even noticed a change going on, but has been merely swept along with the tide of opinion. Standards and values have increasingly been based on personal and subjective inclinations which are relative to each individual's ideas and experiences. This shift in thinking has caused a general drift away from the kind of standards which are thought of as fixed and objective truths, to those which are based upon each individual's personal subjective inclinations and his particular circumstances of the moment.

An example of the influence of this shift in thinking, particularly in the way people are now deciding what is right and wrong, was made clear to me when I was urging a lady to do something which would normally be considered the "right" thing to do. She objected, "But I don't *want* to do it, so surely it would be *wrong* for me!" The argument, simplified, is something like this: what is right and wrong is strictly an individual matter. To be "right" is to be in harmony with your personal feelings and inclinations. It therefore must be "immoral" to allow yourself to be pressured into doing anything which you don't want or like doing.

I call this existentialism. Philosophers might want to argue that this is too broad an application of the term, but it would be difficult to deny that it is some brand of humanism, and it involves a dividing of things into two kinds of "knowing," the objective and the subjective.

This much is clear: there is a great movement going on in which people are being encouraged to believe that the location of value and significance for life ought to be inside each individual. Are you an existentialist without knowing it?

WHY EXISTENTIALISM CAN'T SURVIVE

Existentialism as a philosophical movement is proving to be a passing phase, a stepping stone between world-views, part of an ongoing trend. It's inevitable that this should be the case, although there will doubtless always be those individuals who are going through their own existentialistic phase.

The reason existentialism can't last is because it also has its built-in flaws, just as naturalism has. Remember that an inescapable problem faced by naturalists is their inability to account for the fact that man is rational and that he has a mind that contains real concepts and ideas and makes real decisions. From the view-point of the main tenet of naturalism, (the closed system theory of how nature works), such a mind ought to be impossible, unexplainable, an illusion. Yet it was by using his rational mind that he conceived of and developed the theory of naturalism in the first place! It doesn't work to *think* that thinking is impossible!

Since he couldn't explain this and be consistent with his naturalistic world-view, he took a blind leap, in spite of his objective "evidence" to the contrary, and just declared humanity to be significant and his own rationality to be intact. By an existential leap he became a humanist.

And this is the spot where the inherent weakness in the system shows itself. By relying on two very different ways of "knowing," man found that he was simultaneously both affirming and denying his own rational nature. His "Faith" was like that of the little boy who defined faith as "believing what you know ain't so"! You say this doesn't make sense? That's exactly the point. Existentialism isn't supposed to make "sense," and that's what makes it actually a big step toward mysticism rather than a final solution to the problem.

PSYCHICAL RESEARCH: THINKING ABOUT THE UNTHINKABLE

It is man's nature to be a thinker, to try to make sense out of his observations and experiences. He is always asking questions. This characteristic is especially true of those people who dreamed up naturalism and built this modern world of science and technology. The influence of their Christian background is still strong, and blind faith just isn't good enough. What could be more natural and inevitable than for men to start *investigating psychical phenomena?*

That's exactly what's happening today: there are many scientists who are actively engaged in research aimed at *explaining that which is not supposed to be scientifically explainable!* They're tired of suffering from a philosophically split personality, and are trying to bring together the two parts of their divided world-view. Besides this, the psychical is a very fascinating subject as long as it can be made scientifically and socially acceptable.

One result of research into psychical phenomena is to accelerate the shifting over within psychical-intuitive thinking in the direction of pure mysticism. Of course we don't want to give the impression that this movement is limited to the speculations and theoretical studies of scientists and academics. Interest in mystical subjects is generally spreading throughout our society, and as a style of thinking it has almost taken over.

The increased experimentation with psychical phenomena is of major significance as it has become a trend which will have a big influence on general society. The trend-setters can't be laughed off as just "weirdos," back-room fanatics, or the ignorant and superstitious. They are as likely as not to be found in positions of prominence and influence. And why not? They want to know the answers. It isn't comfortable for anyone to have to leap back and forth from one philosophical foot to the other. No one wants to feel that he must play psychical tricks on himself in order to find peace of mind. No one wants to live with a gap between his head and his heart.

One point we have to remember is that psychical research is a search for powers, places or personalities existing *within* this natural universe. Nature is still believed to be a *closed* system, but just perhaps, the speculations go, there are *other kinds* of forces at work in the universe in addition to the known functions of mechanistic, impersonal, natural cause-and-effect.

The investigations into psychical phenomena mean that theories are being tested. These theories infer that it just might be that the mechanistic and materialistic assumptions of naturalism are false. Is it possible, they ask, that the universe might contain something — or even *be* something — which naturalistic science has failed to recognize? Maybe we need to alter our world-view. Wouldn't it be much nicer or at least more interesting if the psychical side turned out to be a kind of reality which is more than just psychological or imaginary? Suppose that psychical powers and forces not only exist, really and truly, inside us, but suppose there are such powers elsewhere in the universe. Is it conceivable that there really is something or someone out there? — or in there? — or somewhere? — or everywhere?

The circle is in the process of becoming complete. Significance is once again being sought in the invisible powers and personalities which might be surrounding us. Events and experiences are being re-examined on the possibility that events are being influenced by as-yet unidentified entities or forces. Mysticism no longer has to stay underground.

Notes — Chapter Six

1. For discussions of "Determinism" see pp. 24, 33, 38, 39 in:
   SCHAEFFER, Francis A.: *Escape from Reason*; London: Inter-Varsity Fellowship, 1968
2. See C.S. Lewis' *Miracles*, page 17
3. "Hence every theory of the universe which makes the human mind a result of irrational causes is inadmissable, for it would be a proof that there are no such things as proofs. Which is nonsense." Lewis, *Miracles*, page 21
4. John 3:19; Rom. 1:28
5. Webster's Ninth New Collegiate Dictionary defines Psychical:
   > 1. Of or relating to the psyche;
   > 2. lying outside the sphere of physical science or knowledge: immaterial, moral or spiritual in origin or force;
   > 3. sensitive to non-physical or supernatural forces and influences: marked by extraordinary sensitivity, perception or understanding.

   Dictionary definitions reflect the popular usage and common understanding given to words. In this case, the failure to distinguish between supernatural and psychical forces within nature is obvious. If we are to hold to a Biblical world-view, the two terms must be clearly separated in our thinking.
6. "Mannishness" is a term used by Francis Schaeffer to indicate those special characteristics, the image of God, which distinguish man from all other living creatures.

7

It's Nothing New, But Is It True?

WHAT *ISN'T NEW?*

Before going any further, a more precise definition is needed. Just exactly what are we talking about when we speak of "mysticism"?

The term as we are using it designates two separate areas, the one being Satanic and the other psychological.

On the Satanic side are those mystical powers and personalities which exist outside and independent of man, the real forces in this universe which are hostile to both God and man. Mystery surrounds the mystical because it is unnatural, invisible — and inevitably destructive. Being opposed to the truth of God, these mystical beings specialize in lies and deceit. The truth of God is reasonable and rational; they are the ultimate in irrationality.

On the psychological side are the beliefs and ways of thinking within man which agree with those mystical forces, or are most open to their influences. An aspect of man's natural cognitive, or psychological processes, is the *psychical*. Psychical-type thinking is based on non-rational rather than rational processes. Instead of reasoning something out, or applying objective analysis, the psychical thinker acts on intuition, urges and instincts. He doesn't often question his conclusions; he just accepts them.

Mystical thinking takes over where mere intuition and instinct cease. To the out-and-out mystic, the subjective impulses within him are valid and reliable because he believes them to be caused by a higher power of some kind. He has been inspired or motivated to feel or believe as he does. All the study and examination in the world can't analyze the source of such "knowledge," and there's no rational way to explain it to someone who wants "proof." It would probably be true to say that mystical knowledge is believed to come from mysterious sources to the senses — or an internal "sixth sense" — and not through normal learning and reasoning processes.[1]

The obvious question which will spring to mind is, what's the difference between inspiration from a mystical evil spirit, and inspiration from the supernatural Holy Spirit? There are big differences which will be laid out later at the appropriate time, but an orderly sequence of ideas is important.

"THE BEAST THAT WAS..."

Mystics have existed in every age and culture. People tend to think of mystics as being rather mysterious, non-worldly individuals, probably practitioners of esoteric religious arts, having special access into spiritual things. Perhaps they are members of an inner circle of wise men where secret enlightenment is available. You've probably seen cartoons in which the mystic is humorously pictured as sitting in meditation on top of a high mountain or in a cave or secluded in the cell of an ancient monastery.

"... AND IS..."

Actually the world is full of mystics. They might work in the same office as you do, have children in the same school, live in a house just like yours, or be sitting in the pew in front of you at church. Perhaps I'm speaking to a mystic right now. That identity rightly belongs to anyone whose style of believing and thinking corresponds with the assumptions of mysticism. The details might change and the terminology may vary widely, but the principles will fit the pattern. Mysticism is not at all uncommon, especially if we include the large number of people who are being influenced by the trend and are moving in that direction.

The current experimentation with psychical phenomenon is a study in mysticism. Parapsychology is not the same as conventional psychology. The investigation by the modern researcher into psychical phenomena is an excursion into precisely the same kind of powers and personalities which are already assumed to exist by those who hold to a mystical worldview. What this means is that old fashioned mysticism, including the overtly occult, is coming together with modern science and academia.

A popular pretext given by the various branches of science to justify psychical research is that if these powers should actually exist and if we knew enough about them, they would be found to be as "normal" as the more material and measurable components of the universe.

At the same time, overtly mystical practices are increasingly becoming a part of the life-style in the general culture of the Western world. Occultism is enjoying a social rehabilitation. Of course mysticism and occultism have always been around, but for a long time they had been generally shoved into the background. The amazing thing is how quickly such practices have emerged from the dimly-lit room of the spiritist seance, the African hut or the Eastern temple and have become accepted as socially respectable in the "Christian" West.[2]

Once we become consciously aware of this trend rather than thoughtlessly drifting with it, the influence of mysticism seems to be so obvious that we're tempted to see an occultist behind every bush. It certainly is startling to see respectability and credibility being accorded to beliefs and practices which not long ago would have been classified and rejected as occult or pagan. I was astonished — naively, I suppose — a few years ago to discover that the conservative, Midwestern state university in which I was studying, offered a course in witchcraft, with the students organized into a coven. Numerous academics and scientists are studying the "real powers" to be found in traditional religious practices. There's a live issue involved when blatant occultism is considered a proper subject of academic instruction and practice.

Every city has an over-abundance of occult bookstores. Even the more conventional bookstores have a section for the occult, psychical and mystical. Hindu gurus and courses in transcendental meditation are not even novelties anymore. Psychologists are talking about the need to bring the "spiritual dimension" into therapy. By this they don't mean "supernatural" as we've defined it, but the psychical. There are hundreds of well-attended "self-improvement" seminars in which the participants seek to experience a break-through into non-rational levels of consciousness.

IT'S NOTHING NEW BUT IS IT TRUE?

The public media is certainly not exempt from the influence of mysticism. Someone recently stated that the programs on TV which the children all watch are only the modern equivalent of traditional fairy tales. Perhaps so, but just think about those old fairy tales and the origins of so many of them — magic and mysticism straight out of the superstitions of medieval Europe! The difference is that today, the magic is performed by space men or monsters and is all mixed in with pseudo super-science.

We can't blame the kids if they take such myths seriously when there are millions of adults who believe the unfounded and unscientific speculations which say that space just *must* be populated with civilizations so advanced that their science would seem like magic to us.

In many countries in Africa, and probably in other "Third World" nations, "traditional healers" are being officially recognized in hospitals on a professional basis equal to scientific medicine. In religious circles there's a movement to systematically reinterpret and teach "Christian" theology in the light of traditional animistic beliefs.[3]

Have you heard the old adage that there are some people who make things happen, some who watch things happen, and a majority who don't know that anything is happening until it's over? Well, it's happening. Through seminars and symposiums, in periodicals and books, informally and through institutions, at home and abroad, the trend is like a fog quietly thickening and seeping under the doors and through the smallest openings. Certain aspects of it have even been named — the "New Age Movement" — to show that another stage has been reached in human history. Of course it isn't new at all. It's a return back into mysticism. This being the case, we might wonder if a new Dark Age is possibly being ushered in. Some consider it a Golden Age.

". . . AND IT IS TO COME!"

The movement toward mystical ways of believing, thinking and acting is not a brief and passing phenomenon. Society is bound to become saturated with whatever it's soaked in. As with every world-view, the more an individual is surrounded with it, the more deeply it becomes internalized into that person's basic character and personality, his habits and assumptions. It's very important that we clearly understand what's happening, and equally important, that we be able to recognize a leaning toward mysticism, whatever form it takes.

Remember that mysticism is first of all a world-view: it can be recognized on the basis of what is believed to be true and not true, and by how a person comes to understand the things which are significant. If a person holds to a mystical-type interpretation of reality, that person is a mystic even if he's never actually realized it or put it into words. A mystic — or potential mystic — believes that ultimate meaning and value in his personal life depends on what is taking place within his own subjective world, even though he himself may not claim to have had any especially unique psychical experiences. He will be the one who is inclined to ask *who* caused something to occur — who, that is, using invisible psychical powers, or who in the invisible realm of the spirits — rather than first presuming that he lives in a universe which has a basic framework of natural causation.

WHOSE "TRUTH" IS TRUE?

How does this cause you to react? Are you worried? Curious? Do you see the rebirth of mysticism as the hope for world peace and prosperity, and a recognition of long-suppressed truths? Or do you see this as a sign of spiritual renewal running parallel with false teachings? In other words, is the trend basically good or bad?

That's entirely the wrong question to ask. This question is only of value if we look at the issue from a mystical perspective! From the supernaturalist's point-of-view, what an individual happens to think doesn't prove anything. Whether the direction things are going is, in fact, good, bad or indifferent, all depends on *one factor: Which world-view is true?* Popularity doesn't mean validity! A belief or idea is good only if it interprets reality correctly.

How we feel about things is certainly important, but it isn't a sound basis for judgment. The issue is *not* relative to our personal preferences or inner convictions. That would be proving mysticism by mysticism — making truth subjective. It might be that we personally like or dislike one set of beliefs or another. We might want one or the other to be true. We might be absolutely convinced of the validity of our personal convictions. The only fact that really matters, however, is whether or not we have a true and accurate understanding of what genuinely does and doesn't exist — what, in actual fact, causes things to happen as they do. How we feel is important, but the amount of emotion which this subject generates does not make it true. Thus, it is improbable that very many will consider this trend toward mysticism to be irrelevant.

PRODUCING THE EVIDENCE

Is it even possible to know the truth? The proponents of every world-view believe they have good reasons — assuming they believe there is any such thing as reasons — for holding to their particular position. It's important not only to understand *what* is believed, but also *why* it is believed. Obviously it isn't possible to outline all the arguments for and against each point-of-view. It shouldn't even be necessary, because if we once grasp the fundamental concepts and principles involved, the details fall into place as necessary implications. Above everything else then, the issues should be made crystal-clear. All the bits and pieces need to be brought together, and the different ways of thinking ought to be contrasted and compared. In this way we can better come to understand the reasons for believing in one world-view as compared to the others. We'll do this by summarizing and comparing each of the three philosophies, why they believe as they do, and then consider their attitudes toward those who hold other views.

WHY SUPERNATURALISTS BELIEVE AS THEY DO

Supernaturalism, from the Christian viewpoint, is based on the conviction that there is a supernatural God who has revealed Himself through nature, through a verbally inspired book, the Bible, and through His Son, Jesus Christ. The evidences for each of these are believed to be reasonable and open to objective analysis, and the supernaturalist believes that the evidence should be more than adequate if one begins his investigation with a reasonably open mind. The preponderance of evidence, the supernaturalist contends, is that nature is an orderly system, but an open system into which the supernatural realm may, and sometimes does, intrude. Nevertheless, the basic framework of every-day existence is that of natural causation.

To come to a belief in the existence of God and the supernatural realm does not require any mystical enlightenment, but simply asks that a fair examination of the evidence be given. Believing in the existence of the supernatural God is logical because if there is no supernatural, neither is there logic! Our very existence makes sense because man has a rational origin.[4]

The supernaturalist does believe there are real powers and personalities which are mystical, invisible and unnatural forces in this universe, which are not of God but are demonic. Once again however, this doesn't alter the fact that the basic framework of daily life is that of the normaal natural system.

WHY NATURALISTS BELIEVE AS THEY DO

Naturalism appears to be reasonable because of the regularity and order of the natural universe. The evidence, it contends, indicates that nature is a closed system. One looks around and sees only natural things, not supernatural things. Science and technology have proven extremely successful in unraveling the mysteries which once were credited to gods, magic or spirits. As a consequence, it is argued, it only makes sense to conclude that everything without exception potentially can and should be explained naturalistically. The problem, as many naturalists have discovered, is that in actuality the naturalist has great difficulty even explaining himself! Naturalism ultimately finds itself depending on a leap of blind faith rather than on the cold hard evidence it looks for. Of course, when a naturalist does this, he isn't truly a naturalist any longer but has actually become an existentialist of sorts.[5]

WHY EXISTENTIALISTS "BELIEVE" AS THEY DO

Existentialism does not legitimately have any grounds to prove nor much of anything to argue, except that faith is unarguable. It simply says in a wide variety of ways, "Believe with your heart what your head rejects." Because of its philosophical schizophrenia, it allows one to be, so-to-speak, a "Sunday-go-to-meetin' Naturalist." The rational side is a naturalist, and the irrational side is a "believer." But when an existentialist attempts to argue his case, he is not really an existentialist any longer. Depending on what argument is used, he becomes either a supernaturalist, a naturalist or a mystic.

WHY MYSTICS BELIEVE AS THEY DO

Mysticism founds its position on both subjective and objective evidence. On the one hand the mystic will say, "I know that what I've personally experienced is real but it is beyond description. You'll have to experience it for yourself." Seminars, sessions, seances or circles may be conducted in the hope of making these experiences possible.

On the other hand, the mystic may produce eye witnesses, photographs, tape recording or books in which are recorded great numbers of objects and events. "All these," they argue, "are not explainable on the basis of any know laws of nature. They prove that there are powers — or personalities — which go beyond the impersonal and mechanistic functioning of material things."

WHAT MYSTICS BELIEVE

How mystics believe, and *why* they believe as they do have been considered. Now it is time to consider some examples of the more popular doctrines and tenets of various mystical groups. Some mystics will believe that the various unnatural occurrences are the doings of the spirits of the dead, or perhaps come from more highly-evolved extra-terrestrials — beings from outer space. These personalities or powers are believed to explain such mysterious and controversial things as psychical healings, extra-sensory perception, precognition, telepathy, telekinesis and the like. Many will believe that there are basic laws of nature which have not yet been discovered or understood but which will be uncovered when we have sufficient information.

There are those who hold that all living beings contain a part of a single great Life-force or energy, and this energizing power forms their definition of "God." A clue that someone holds this pantheistic view is the often-heard, "God is in all of us!" Still others hold that the universe itself — and if not the universe, then this planet — is one vast living entity (the buzz-word is "holistic") with which we must get in tune (the buzz-word is "ecological"). A lady was recently heard in a radio interview referring to the earth as "Planet." Not "the planet" but using the term as if it were the proper name of a living person.

In every case, mystics will be found to believe that this life is only understandable and only takes on true meaning when one's subjective life experiences are in communication with certain invisible personalities, or in unity with invisible powers which are present and active all around us.

While some mystics classify themselves as Christians and credit the powers to God, many others consider the powers to be impersonal forces or perhaps, spirit beings. This should make it obvious that very similar-appearing phenomena could presumably originate due either to supernatural intervention or mystical powers.

Satan worship, along with every other occult practice and belief, is obviously the epitome of mysticism. All occultism is mysticism, but we can't accurately say that all mysticism is occult. Some mysticism is psychological only. But although the identity of the powers may be debated, from one extreme of mysticism to the other, the basic principle is the same. The universe is conceived of as a mixture of material and immaterial things, the material held together, manipulated and given meaning by mystical forces, whatever they might be. In the final analysis, the mystic doesn't actually believe in either supernature or nature, but in one system with no clear division of the natural and the supernatural.[6]

HOW WORLD-VIEWS INFLUENCE OUR ATTITUDES

How do these world-views affect the attitudes of those who

hold them? What a person believes is a fairly reliable way of predicting how that person may be expected to look on those who hold to other belief systems. Agreeing with one another on matters of fact is important, because trust and confidence and respect for another's judgment is closely related to the degree to which we agree.

While taking into consideration the usual human inconsistencies and ambivalence, here are some clues to attitudes which can be expected between the proponents of each of these world-views:

Classical naturalists, those few that still survive, will doubtless be found to look with disdain upon both supernaturalists and mystics as being ignorant, superstitious or stupid. Existentialists can be tolerated since they're not challenging the assumptions of naturalism.

The existentialist, basing his life on a subjective and non-communicable blind faith, is in no position to consistently object to what any other person believes. But we need to remember that his attitude will be affected by the fact that the existentialist is likely to be a disillusioned naturalist on his way toward becoming a mystic! He can't be expected to be consistent. His biggest objection is toward the supernaturalist who insists that there are objective grounds for belief.

The mystic will object to supernaturalism as being both too naturalistic and at the same time so unnecessarily other-worldly that it is blind to the very real powers which are right here. Both the supernaturalist and the naturalist are held to be too stubbornly materialistic to admit that the universe is undoubtedly permeated with powers which are beyond the natural. It appears to be difficult for most mystics to resist the temptation to flaunt their "inside" information or experience. The mystic may be very secular, but he is at the same time fervently "religious" and usually evangelistic toward others. He is, after all, the ex-existentialist who has had his eyes opened!

We supernaturalists are probably the least consistent of the lot. We react negatively to the naturalist because of his anti-supernaturalism and rejection of the miraculous, and yet never expect or believe that a real miracle might actually take place when we've prayed about something. Then again, the average man in the pew will probably have, at best, only a foggy idea of the difference between supernaturalism and mysticism.

It isn't so much mysticism itself, as it is *certain types and kinds* of mystics to which the average supernaturalist is opposed. Many will be hesitant to object to anything which is supposed to be "spiritual" because it must be much on the same side we are. Having been influenced by psychical-type thinking, being sincere — a subjective state — overcomes much factual error. An exception to this attitude would of course be something clearly labeled spiritualistic. occultic or demonic. Theoretically, Christian supernaturalists have always believed in the reality of Satanic mysticism, but tend to be dubious about attributing a specific event to an on-the-spot demon. It's easier to accept demons in the Bible than in the board room or bedroom.

One school of thought among Christians views the lack of mysticism in the church as a serious failing which ought to be — and is now being — rectified. We need, they say, to reclaim some of the mysticism left behind in the medieval church. *Both mysticism and supernaturalism are good and are not incompatible.* This claim is too important and too common to be ignored.

MYSTICISM IN THE CHURCH: SUPERNATURAL, SUPERSTITION OR SATAN?

Would the supernatural world-view actually be benefited by

adding a bit of Godly mysticism — if there is such a thing? Is it possible for mystical beliefs and practices to be divided into good ones and bad ones? If so, how can we tell the difference? By attempting to clearly distinguish between the supernatural and the mystical, are we not in danger of being "unspiritual"? Surely we want anything which draws us closer to the Lord, but for that to be the case, it must be truly from God and not be a spiritual counterfeit, a delusion or a diversion. The problem with the claim that Christians need some of each, is that it can only be true from a mystic's view-point: based upon the way mystics decide what is true and what is not. From the supernaturalist's position, if one of these ways of thinking is true, the other is at best, untrustworthy and at worst, unspiritual. We'll have to check this out carefully. Truth is the issue.

Now we're getting down to the "bottom line" of the problem. The church, being influenced along with the rest of society, is sliding strongly and significantly out of supernaturalism and into mysticism. We've previously examined the reason naturalism has gone unnatural and into its movement in the direction of mysticism. It's evident — and the evidence will be examined in much more detail later on — that this same thing is happening to the church as well. Why? There has to be a reason. To say that the implications of this event are momentous is an extreme understatement! What is the truth? Is the trend of the times genuinely taking us into a new age of mysticism? What is truly happening, and what is truly right? What *is* and what *ought* to be? Is it legitimate and pleasing to God?

No, God isn't pleased! The crucial battle in this cosmic war may be turning on this issue! Ideas do matter, very much. That which is at stake is the very nature of the Faith. There *is* a sharp distinction between these two world-views, and the practical implications are of the greatest significance. They are vastly different as philosophies of life.

But in spite of great differences, supernaturalism and mysticism are constantly being confused with one another, and there's a reason for such confusion. First of all, because it's easy to assume that whatever isn't natural must therefore be *supernatural*. The mystical realm as a source of real powers and personalities antagonistic to God is too lightly considered. ("It could never happen to *me*!") Secondly, mystical-type thinking is easily assumed to be only deep concentration, or as divine intervention into a person's mind and emotions. Thirdly, every person spontaneously recognizes and accommodates himself to all three main areas of interest and emphasis. In agreement with naturalism, it's true that there *is* a natural universe containing concrete particular objects which function predictably. In harmony with supernaturalism, man *does* think rational thoughts and make genuine choices. As mysticism insists, there *is* a psychical part to human nature as well as heavy evidence for psychical forces outside of man. Yet when these three major components of our life experiences are placed in differing orders of priority and given differing emphases, the consequences are dramatically different, incompatibly different. In other words, which one of the three is the big factor, the one that takes priority over the others as that which gives meaning and value to life? What relationship do these components have to one another? These questions will be dealt with in some detail, especially in Chapter Eleven.

Different world-views — with shadings and stages of transition from one to the other, and with lots of confusion and inconsistencies in between. Each has its own conclusion as to what life is and how to live it. Each has a conviction regarding what is true and significant.

From this point on our concern must be this: just how far

toward mysticism has a once-supernaturalistic Christianity slid, and in what specific areas of Christian faith and practice can this be seen, and what difference does it actually make?

Notes — Chapter Seven

1. Jude 10
2. Gabriel Setiloane, Associate Professor in the Department of Religious Studies at the University of Capetown, South Africa, says that Western man has "brainwashed" himself with "all kinds of sophistication which he calls Civilization, Christianity, Science or Philosophy" to the point that he is impervious and insensitive to Divinity — the "Vital Force" which is not a Personal God, but which is more true to reality than the traditional "Western" God. He continues:

"That is a grim picture. Two lights of hope relieve its horror: first the so-called 'Occult Cults' so ubiquitous in Europe. . . . Are they not some remnants of the 'Old Religion' of Europe, their 'Primal Vision' only spoiled and contaminated by the barbarities and accretions developed in the dark vaults to which they had been condemned by the Inquisition, the Enlightenment and Civilization.

"The second window of hope is (that) . . . Western man seems to be stumbling on the inevitable Reality of . . . The Source of Being. . . . Human Behavioural Sciences, Psychology and Psychiatry, as well as Biology are other areas where modern science has come to confirm the 'Primal Vision' of Africa."

SETILOANE, Gabriel M.: *African Theology, An Introduction*; Johannesburg: Skotaville Publishers, 1986

3. The book cited in the previous reference is an example. Setiloane has written especially for African youth to verbalize and systematize traditional religious beliefs. He holds these to be greatly superior to orthodox Christian teaching. There is an interesting correlation between his position and that of Liberal or Neo-orthodox theology in the West, and he cites such sources in a favorable light.

4. I Cor. 1:18-29 Numerous excellent books on the subject of Christian Evidences, or "Apologetics," are available for study, including the following:

CHAPMAN, Colin: *The Case For Christianity*; Grand Rapids: Wm. B. Eerdmans Publishing Co., 1981

GEISLER, Norman: *Christian Apologetics*; Grand Rapids: Baker Book House, 1976

LEWIS, C.S.: *Miracles, A Preliminary Study*; New York: The MacMillan Co., 1947

_____: *Mere Christianity*; London: Collins, 1952, 1973

LITTLE, Paul: *Know Why You Believe*; Downers Grove, IL: Inter-Varsity Press, 1971

MCDOWELL, Josh: *Evidence That Demands a Verdict*, Rev. Ed.; San Bernardino: Here's Life Publishers, 1979

_____: *More Evidence That Demands a Verdict*; San Bernardino: Campus Crusade For Christ, 1975

SCHAEFFER, Francis A.: *The God Who Is There*; Chicago: Inter-Varsity Press, 1968

_____: *Escape From Reason*; London: Inter-Varsity Press, 1968

_____: *He Is There and He Is Not Silent*; London: Hodder and Stoughton, 1972

5. Whitehead, a generation ago, commented in a famous series of lectures, that the mechanistic explanation of nature was efficient, but made nature a dull and meaningless affair. However, the concept was "... not only reigning, but it is without a rival. And yet — it is quite unbelievable." Pg. 54

WHITEHEAD, Alfred North: *Science and the Modern World*; New York: The Free Press, 1925, 1967

6. Notice how this holistic view contrasts with the Supernatural/Natural distinction of the Biblical world-view (Chapter Four) but coincides with the medieval mind (Chapter Three).

8

"The Creature Rather Than The Creator"

Remember the Biblical scenario from Chapter One? Satan was cast out of Heaven because he wanted to set up his own independent kingdom. Do you think he gave up that ambition after he lost the Heavenly war and was cast out into this universe? Not at all! In effect, he must have said something like this:

"Well, if I can't have a kingdom there, I'll have this one!"

It became his goal to change the entire belief-system of this universe's inhabitants. Step-by-step it could be done — *if!* If the door between Heaven and earth could be kept closed, if the communication between the two realms could be lessened, if man could be convinced that he himself could be like God! Above everything else, if man could just be convinced to *act as if Heaven does not exist!*

MYSTICISM: A SATANIC CONSPIRACY

How to go about it? What did he, Satan, have in his favor? Here he was along with all his followers, evil spirits exiled in the natural universe. He was not even constructed of the same "stuff" as was to be found in nature.

That's it! There's the answer! He and his cohorts were all *spirits*, which simply meant that they were composed of the material of the supernatural universe and not therefore fully subject to the limitations and laws which applied to the material stuff of this natural universe. To natural man, they and their powers would be abnormal, unnatural, super-natural, mysterious. This offered all sorts of possibilities.[1]

Using his unearthly powers, he could claim to be a messenger from God, or even to be God Himself, and get people to worship him. The potentials to be found in idolatry, witchcraft, paganism and even organized Satanism all opened up before him.

He and his cohorts could conceal their presence and then influence men's minds by planting ideas which they would think were their own. They could tell humans that there isn't any supernatural; nature is everything. Or they could create a multitude of other philosophies just by taking advantage of certain aspects of human nature.

He could mystify men with fantastic powers and lead them astray with the hint that perhaps these powers could be their very own. Let them believe that they could manipulate his powers. He must keep them thinking only about the forces and the faces which are found in this universe and keep their mind off God! Keep it mysterious. Keep them curious. Keep them wanting more. The bait must taste good!

Satan is still at it today, only he has thousands of years worth of infiltration, experience, organization, momentum and degenerated human nature working for him. Now he can lay claim to precedent; time seems to make anything legitimate. He's a practiced liar, thief and murderer. We aren't entirely ignorant of his devices. During all this time he has been substituting mystical concepts and assumptions in place of supernatural truths, counterfeiting that which is from God with his own brand of religion, and luring everyone possible into slavery to sensuality.

CORRUPTING THE CHRISTIAN VOCABULARY

That which comes from God is supernatural, not mystical. If it's from God, even though we inaccurately call it "mystical," it's still good; however, it has the wrong label attached. If we label something as "supernatural," and as being from God when in fact it's mystical in origin, that label doesn't make it legitimate. That's one good reason for making certain that our terminology is accurate, and why in this book every effort is made to use the term "mystical" with a definition which is precisely distinct from the things of God and His supernatural realm.

The primary reason we need to get our terminology straight is because we must get our facts straight. In order for words to communicate ideas accurately, their meanings have to be understood correctly. The reverse is also true: mixed-up thinking leads to mixed-up word-meanings.[2] Powerfully influencing the slide of the church toward an unscriptural and ungodly mysticism are the changes in definitions which are being given to certain key words in the Christian vocabulary. Inevitably these perverted word meanings result in perversions of concepts and ideas as well. In each case the change is from a definition which has supernaturalistic significance to one which is psychical or mystical. Vital Biblical concepts are being twisted around so that they end up taking entirely different shapes and thought-forms, and are made of totally different mental stuff than was previously the case. Many Biblical practices and supernaturalistic ideas are being incorrectly re-interpreted into mystical concepts and styles of thinking. In some cases, key words have been entirely re-defined.

The issue is much greater than whether our faith is to be both objective *and* subjective. Two entirely different ways of thinking and reasoning are involved, and as a consequence, the very definition of what it is to be a Christian is being questioned! The church is being assisted into mysticism.

A NEW MYSTICAL DEFINITION OF "FAITH"

Faith, as it has come to be understood by countless Christians, no longer means exactly the same thing as it did when our New Testament was written. To "have faith" now often tends to bring to mind distorted ideas and assumptions which more closely coincide with a mystical interpretation of reality rather than the supernatural world-view.

Faith wasn't difficult to understand when the Apostle Paul commanded the Jailer in Philippi to believe.[3] By the way, in the language of the New Testament, "faith" and "belief" are simply different forms of a single root word, varied to fit the grammar or context. The words were commonly used in the every-day vocabulary of people all around the Mediterranean world. They knew that to believe, to have faith, meant to face up to the reality and validity of certain facts, and to live in accordance with those facts, to make a commitment of some kind. One with faith is one who trusts in, depends upon, adheres to something or someone outside of himself.[4] Acting on faith was simply a matter of acknowledging the validity of the evidence and altering one's life to fit those facts. Faith was doing the sensible thing.[5]

But faith has increasingly come to mean something far different than commitment to certain facts, the surrender of the mind and will to reality. Faith has moved into the mystical realm whenever we hear someone answer a question by saying, "Well, that's just one of those things we must accept on faith!" Mystical faith is irrational.[6]

This is not genuine Christian faith, but rather *existentialism*, which makes a leap of blind faith. It is not *the* Christian faith that relies on irrational knowledge. The Biblical, Christian faith has a foundation of sound evidence which can be communicated; mysticism, on the other hand, is not only unprovable, but is ultimately non-communicable, because it is not based upon objective evidence but upon subjective experience. True faith is a matter of fact, not a matter of intensity. The supernaturalist's faith and the mystic's faith are distinguished by the kind of "knowing" which is required.

To crystalize these distinctions we need only to contrast the mental framework of the supernaturalist's world-view with that of the mystic. The genuine Christian faith depends entirely on the reality and historicity of Jesus Christ for meaning. Mystical "faith" depends on the strength or quality of a deeply subjective attribute residing within the believer himself. Christian faith is faith in Christ, who previously was here but has now entered into the supernatural realm, someday to return. Thus it is supernatural faith. Mystical faith, in the final analysis, is faith in Faith, an invisible and mysterious force working inside the "believer," hence is mystical. The Christian trusts Christ because He has demonstrated Himself utterly trustworthy.[7] The mystic must depend on the sincerity, strength and intensity of his own attitudes and emotions.

The next time we catch ourselves believing something to be true on the basis of intuition or gut-level conviction, we ought to have the good grace and good sense to call it what it is — human wisdom, perhaps valid, perhaps not, but still human. It must be tested before it is trusted. First come the facts and *then* comes the faith, with feelings developing as a consequence.

DON'T CALL IT "MYSTICAL," CALL IT "SPIRITUAL"

Closely related to this distortion of the true meaning of "Faith," is the mystical perversion which has also been taking place in the distinction between what is *spiritual* and what is *carnal*. The Bible often makes such a distinction.[8] To have the "mind of the Spirit" rather than the "mind of the flesh" is of grave importance to every person who wants to please God. On the one hand is the Spiritual mind; on the other is the natural, earthly, fleshly or carnal mind. We are called to "set" our minds on the former, not on the latter. The Spiritual mind-set is said to be capable of understanding the things of God, while the carnal mind-set is incapable of such understanding.

What is it to have a Spiritual mind-set? Christian preachers and teachers exhort us to have it. Some Christians accuse others of not having the mind of the Spirit. Sometimes we get carried away on guilt trips because we feel that we must be thinking and functioning with our own fleshly mind rather than having the mind of Christ, the mind of the Spirit, at work in us.

Understanding the supernaturalist world-view makes the identity of the mind which is set on the Spirit both straight-forward and evident. The realm of the Spirit, Heaven, is the realm of God, who is a Spirit.[9] The natural realm is this universe. The Spiritual mind is constantly aware of God and of His Son who sits at His right hand. The spiritually-minded person is conscious of the reality of the supernatural universe and makes his judgments in that light. To him or her, this is simply a matter of fact, but it is such an overwhelmingly important fact that everything in this universe must take it into constant consideration. The Spiritual mind is objective, realistic, and practical. There is no way to have a spiritual mind-set except by believing in the objective existence of that other universe. Therefore the major characteristic of a spiritually-minded individual is that he is "other-minded." His daily experiences are evaluated in light of the reality and superiority of the supernatural God. Rather than this causing him to be so "heavenly minded that he is of no earthly good," this way of thinking is absolutely practical because it fits with the facts of reality as they truly are, rather than disregarding more than half of what is! Remember that it was the supernaturalists whose world-view resulted in the way of thinking which produced our modern scientific age.

The spiritual mind is spiritual because of *what it thinks*, not because of *how it feels*! It is the mind which is fixed on — which strives for and seeks the interests of — God's will and pleasure, rather than those of this earth. The spiritual man is not looking for security, meaning, or godliness inside himself but rather outside, and not only outside of himself, but outside of this natural realm entirely.[10] He has an objective standards by which he can evaluate each event, object, concept or experience. Because of this fact he is able to live a sensible, stable and realistic life. He has no difficulty in living with the idea of nature because of his belief in a separate and distinct supernatural realm. He can accept the reality of a universe functioning on the principles of "mechanistic" and normal causation because he doesn't attempt to find ultimate value and meaning within the boundaries of the natural universe.[11]

Heart-felt conviction, intensity of devotion, elation and excitement over the things of God are not, in themselves, the "mind of the Spirit" but are the internalized reactions which ought to be the spontaneous results of saturating one's mind with the facts revealed to men by God. They are psychical consequences of having the mind of the Spirit.

It is not the spiritual mind, but the worldly and natural mind which lays claim to *irrational* knowledge. The appeal to irrational knowledge is the hallmark of mysticism. The mystical realm is the realm of Satan. His realm isn't in Heaven. It is entirely in this universe, in nature. The natural man looks for meaning in the powers, places, or personalities which are *here*. His search for value and meaning is satisfied only if he finds evidence that invisible or paranormal powers are at work behind the forces of nature or inside himself.

The carnal man is *sensual* rather than spiritual because of his appeal to the senses.[12] It ought to be remembered that mystics make two claims to justify their beliefs: first, there is the appeal to objective evidence — that events have genuinely happened which can't be explained by any of the laws of nature. Then there is the claim to having a knowledge of things which can only be gained by personal experience. Naturally an individual's personal and internal experiences can neither be proven nor disproven. It's rather like having the doctor tell a patient that nothing is wrong, and having the patient argue, "But doctor, it *hurts*! You can't feel it but I can!"

This claim to possess subjective evidence is literally claiming to have "inside knowledge." It is irrational knowledge, or at least non-rational, because it isn't based on the logical thinking of the brain which God gave man, the examination of the evidence, nor on objective verbal communication. What could be more carnal, sensual, earthly than to set one's mind on the feelings, urges, impulses, intuitions or sensations which reside within one's self?

I once lectured before a group of Bible translators on the

influence which differing world-views have had on the ways of thinking and reasoning of various cultures. When the discussion was opened up for questions, one lady stated that the lecture troubled her:

"Let me see if I understand you. You seem to be saying that you believe there are three powers working in the world: God, Satan and natural cause-and-effect."

"Yes," I answered, "That's more-or-less what I'm saying."

"Oh, no! I can't believe that!" Her reaction was both indignant and emotional. As it turned out it was natural cause-and-effect which she was rejecting. To accept that she lived and acted in a world where the normal framework of events was based on the orderly but "impersonal" natural system was, to her, a denial of the spiritual. All good events are caused directly by God; bad things are from the Devil. Nothing, theoretically, was due to the fact that we live in a system which has a framework of natural causation. Hers was a mystical world-view through-and-through with all experience being interpreted on the assumption of personal, yet invisible powers being the key to understanding.

Satan has done a switch on us! He's taken the *mystical* and renamed it the *spiritual*, and then accused the supernaturalist of being "unspiritual" and guilty of relying on the mind of the flesh. "Heart knowledge" must be allowed priority over "head knowledge," the sensual over the sensible, the irrational over the rational!

To blame the drift of the church toward mysticism all on the devil would be, itself, a mystical way of thinking. Man can get himself in plenty of trouble on his own. We dare not overlook the characteristic of fallen human nature to prefer the sensual to the sensible, since feeling good is more comfortable and pleasant than tough thinking.[13] Many other factors have contributed to the pressure on the church to drift toward mysticism. The constant influence of the world around pushes us in the wrong direction. So-called "Christian Existentialism" has played its part with trends in theology, and this has tended to trickle down to the level of the pulpit and pew.

The bottom line is a conflict over whether or not Christians have a right to trust their heads before they trust their hearts. *Thinking* is the issue and our world-view determines both *what* we think and *how* we think. Is thinking a sin? The mystic says we can know spiritual truth without going through the process of logical thought. Next we're going to think about how we're meant to do our thinking.

Notes — Chapter Eight

1. Eph. 6:12; II Thess. 2:9-12
2. Gal. 1:6,7; II Tim. 1:13; II Pet. 3:15-17
3. Acts 16:31-34
4. Rom. 10:17; Heb. 11:1
5. James 2:18
6. We must be careful about the ideas which we have in mind when we use such terms as "reasonable" and "rational." True Christian faith is never unreasonable even when surpassing the limits of human reason. Faith does not go beyond the rational; "irrational" infers that which is contrary to, not simply beyond, reason.
7. II Cor. 3:5; II Tim. 1:12; Heb. 3:1-6
8. Rom. 8:1-17; I Cor. 2:2,14ff., 3:1-3; Gal. 5:16-25; Eph. 2:1-3; Phil. 3:19; Col. 3:1,2; II Pet. 2:1-3,10,18
9. John 4:24; II Cor. 3:17; I Tim. 1:17
10. I Cor. 15:25-49; Phil. 3:3; Col. 2:16-23; James 3:13-18
11. Gen. 8:22-9:3; Psa. 19:1-6; Eccl. 1:1-11; Rom. 8:18-25
12. I Tim. 5:11; II Tim. 3:1-7; II Pet. 2:2,18; I John 2:16,17. "Sensuality" need not be limited to sexuality. The issue is the placing of priority on feelings and experiences as contrasted to concepts and principles.
13. Eccl. 8:11, 9:3; Jer. 17:9; John 3:19-21; II Pet. 2:10-12

9

"As A Man Thinketh In His Heart . . ."

A wise old man used to repeatedly say to us, "Think your way out if it!" Right thinking, hard work under the best of conditions, is made harder still by the fact that as mysticism has been increasing its hold, so has the mystical "logic" and style of thinking. There is a growing assumption that knowledge on a non-rational level is somehow deeper and more profound than conceptual or analytical thinking.

Every world-view encourages, and makes almost necessary, its own unique way of reasoning, fostering its own attitude toward the kind of thinking which is valid and acceptable. What we believe about what exists and what doesn't exist, and how we explain phenomena, determines the difference between what is reasonable, as compared to what is ridiculous and perhaps entirely unthinkable.

THE NEW TREND IN "CHRISTIAN" THINKING-STYLE

The church is losing its mind! The supernaturalists' priority on objective, conceptual, analytical thinking is becoming increasingly unpopular, and psychical thinking is being falsely equated with the thoughts of God. The thinking-style which found strong encouragement as a result of the restored Biblical world-view must not be allowed to become unthinkable. We must think our way out of the slide into mysticism.

Re-restoring the supernatural world-view is made vastly more difficult when the supernaturalist's logic is also being phased out. The church ought to possess a deep concern not only to be "*spiritually* minded," but to be "spiritually *minded*." Satan, liar and deceiver that he is, has done the same thing with "spiritual thinking" as he has with "faith" and "spirituality." What was once considered to be proper and sound reasoning is now being labeled "unspiritual" and "fleshly." Actually such a change-over in thinking style is inevitable when a change in beliefs about reality takes place.

The subtle influence of mystical-type thinking has been a pervasive, corrosive undercurrent slipping into the general culture for a long time. It would be most unlikely if any one of us were to have escaped entirely unscathed. Not only must we check out and seriously re-think our world-view to purify it from the influence of mysticism; the orderly thinking, the logic, the style of reasoning which is increasingly being neglected or rejected must also be restored. If the slide into the mystical is to be reversed, it will be necessary to *unthink* some mystical-trend ideas, restore Scriptural definitions to words, and correct basic assumptions in both the facts that we believe and the way we argue and reason.

THE NEW "SPIRITUALITY" AND IRRATIONALISM

Now the going really gets tough! To be exposed as inaccurate is bad enough, but to be exposed as irrational is something else, especially if being irrational also labels us as "unspiritual"! Nevertheless, having the mind of the spirit or the mind of the flesh is too vitally important to base on unwarranted assumptions. Each step that we take from this point on might be so emotionally-laden that it will be like walking on a mine-field. Carefully and prayerfully, we must think our way through it!

The clash between the supernaturalist and the mystical ways of reasoning and thinking has created a conflict over the proper way for a Christian to know things, to make judgments and decisions. Doctrinal differences also become immensely magnified as a result of the vast differences in the ways of thinking. The separate segments among evangelical churches generally *claim* to believe nothing radically new, and in fact exert considerable effort to demonstrate that their particular position has the weight of historic and scholastic precedent in its favor. But major changes are undeniably taking place, and a great deal of this can be traced directly to the slide of the church back toward a mystical style thinking.

Personal experience and observation should make it possible for most individuals to figure it out for themselves. Each major movement of the church, each congregation, each fellowship group, each individual Christian for that matter, always leaves a special impression on others. A unique style, personality, preoccupation, consuming passion, mind-set — call it what you will — is developed as a direct result of the particular things which we most often think about and the way that we think about them. Any specific group invests time, energy and emotions in those areas which are believed to be most important or significant.

Check out what's happening: What are the oft-repeated topics of conversation? How are decisions made? How are problems solved? In fact, what are seen as being problems? If somebody makes a statement of fact, what is it that makes that statement authoritative or truly factual? What is the main impression that we come away with? Even a limited amount of experience will prove that massive differences exist in these areas.

An important example of how this affects the church is in the lack of practical Christian unity. Divisions are as likely to come as a consequence of differences in ways of thinking and feeling, as in technical differences in doctrine. Facts on which we supposedly agree, certainly do not get "equal time" in our teaching. Whatever might be the theoretical similarities in doctrine, the broad general impression of the things done and said comes across far from identical.

The point is, there is closer *technical* agreement in doctrine than there is in actual unity of faith and practice within the church. Differences in ways of reasoning are even greater than the differences in doctrinal beliefs which are segmenting the church. We don't just believe differently; we think differently.

THE CLASH BETWEEN **WHAT** *WE BELIEVE AND* **HOW** *WE BELIEVE*

This hits at the very heart of the problem, the nature of the slide of the church into mysticism. Supernaturalism is built on the factuality of God, His nature, character and deeds as revealed to us through Jesus Christ in this natural universe.[1] *What* is believed is of prime importance. The value of faith is determined by the trustworthiness of the Object of faith. Truth is first and foremost, *objective*. *How* the supernaturalist spontaneously tends to think flows from this position. "Thinking" and "knowledge" are assumed to be synonymous with intellectual activity.

Mysticism on the other hand — including the mysticism of

the church — places more stress on *how* one believes than on what one believes. Thinking as an intellectual exercise is not encouraged; true spiritual "knowledge" comes at a deeper level than the intellect. Therefore possessing the truth depends highly on one's *subjective* state. Prominence or emphasis in preaching and practice leans heavily toward the need to be intense and sincere, toward a definition which regards faith as a mystical quality within. The longer this is stressed, the stronger the trend will become to move from psychical-intuitive, into an ever more mystical emphasis.

Because the church is caught between supernatural ways of reasoning and mystical ways of reasoning, two contradictory methods of making decisions and judgments are in simultaneous use among Christians. The thinking processes which we use occur so spontaneously and quickly that it isn't usually noticed that a "procedure" or method is being followed. We rarely think about how we think!

THE THREE COMPONENTS OF "KNOWING"

In order to become aware of what is happening in our minds, let's consider how human thinking processes work. How do we know what we know?

There are three basic facets to "knowing": the *perceptual*, the *conceptual* and the *psychical*, three ideas or terms with which we've already become familiar. It might help us in organizing our thoughts to notice that these three aspects of our mentalities are somewhat related to the three great world-views which have been under consideration earlier: the naturalistic, the supernaturalistic and the mystical. Each world-view has tended, in principle, to pay more attention to one of these three ahead of the other two, as being of greatest significance.[2]

THE PERCEPTION

Perception is the most obvious, the awareness of "concrete particulars": the specifics, the details of this natural universe which surround us. This knowledge comes to us directly by observation or through our other senses.

All of us perceive, and do so in the same manner. Regardless of culture or race, our brains are designed in the same fashion. The same five senses, neither more nor less, are used by every psychically whole person to pick up and register the objects and events which surround us. It has been estimated that our brains register an average of more than ten-thousand separate bits of information each second. Praise God for the wonderful gift of the brains He's given us! Isn't it a sin not to use them to greater advantage than we normally do?

But having brains which function in such marvelous ways does not mean that we therefore interpret all this input in anything like an identical manner. The problem lies in the fact that we don't all use, evaluate and arrange our perceptions in the same way.

The Positivists, (a certain "school" of naturalistic scientists), once had the idea that they could objectively investigate and experiment with the natural phenomena around them, uncovering the natural laws and principles which were associated with them This would ultimately allow them to unravel all the secrets of the universe. It took a while to realize that all observations and conclusions are *preceded* by assumptions about how we observe and know. We *look* at the same things but don't *see* the same significance or draw the same conclusions.

We perceive such a wide variety of things that we're only able to pay attention to a small percentage of the information which our minds take in. We must pick and choose, sorting out the bits and pieces which look significant. Everyone practices selective perception. The difference is in that which is selected as significant, and that which is passed over as trivial.

CONCEPTIONS

In spite of the fact that our *perceptions* may be registering the same things, our *conceptions* about this information will differ dramatically. Because we have different ways of *believing*, we have as a result, different ways of *knowing*. The details and specifics of our experiences — the concrete particulars — only have meaning in the light of the ideas which we already have about them. This is where the second component of man's knowing, *Conceptualization*, enters the scene.

Concepts are the abstract mental pictures which we conceive as being true or significant in regard to particular items or events. Laws, for example, are concepts, as are moral principles, generalities and universal truths. Concepts are real, but a different kind of reality than the material objects which our senses record. Concepts and ideas are real on a mental level. We analyze, compare, and categorize; we classify and group all the separate particular things as we relate them to other things. Each and every item only has meaning when we attribute certain characteristics to it. True knowledge and understanding are more than the accumulation of detailed information which comes from our perceptions; but are related to ideas. As perception deals with things, conception deals with ideas about things. Thinking conceptually, we ask such questions as, "How do these details fit into the greater plan? Where do they fit in the pattern?" Until such a time as we do this, any particular detail doesn't make sense, it just *is*.

Conceptual thinking places the emphasis on head-level functioning. It seeks, above everything else, to be rationally sound and consistent. Conceptual thinking is based toward systematizing and organizing ideas on the basis of the principles and generalizations which apply. A high value is placed on analytical and objective methods of reasoning. It is concerned with the influence of history and with classical logic. When we place priority on the conceptual we will be interested both in the big picture of over-all principles and generalities, and also in the step-by-step process of methodically tracing through all the implications of an idea.

Notice again the relationship between the supernaturalist world-view and conceptual thinking skills. Looking at the universe as orderly, systematic, and reasonable, causes a person to expect his experiences to be consistent with the cause-and-effect framework. He is more likely to think of the consequences of his behavior. He projects his mind into the future in an attempt to predict the outcome of events. He learns to figure out the details for himself once he has grasped the basic concepts which underlie them. Working on the assumptions of supernaturalism ought to improve one's conceptual thinking skills.

To the supernaturalist, conceptual thinking is essential if one is to grow in his understanding of supernatural things. Perception doesn't give us enough information. God must tell us, so God's revealing of Himself through His Son and His written Word are essential additions to His revelation through nature. As a result, man can think accurate thought and know true things about God's character, His nature, His Heavenly universe and His will.[3]

THE PSYCHICAL

But who ever went through life carefully observing even the major items in his experience, then consciously analyzing and organizing them systematically and drawing significant conclusions based on carefully-worded presuppositions? Nobody, that's who! As we grow up and accumulate a store of experiences and memories, and after we've absorbed a particular world-view (inconsistent creatures that we are), our minds often draw conclusions so quickly and spontaneously that no conscious process is necessary or possible. Our past experiences, pleasant and unpleasant, have also conditioned us (creatures of habit that we are) to respond emotionally and instinctively.

This third component of our knowing processes is the *psychical*. Part of the nature of all human "knowing" in this aspect. The subjective element, as well as the objective element, is in the nature of every man. Who isn't aware of the powerful urges and inclinations which so strongly influence our attitudes and actions? Thoughts and words and ideas come unbidden: a conviction that certain things just *might* be right — or wrong — although we might not be able to give any logical justification for how we know or why we feel as we do.

For example, suppose you were to walk into my office and find — perceive — hanging on the walls, several hunting trophies — the heads, hides or horns of various African game animals. How would you find yourself reacting to this bit of information? While it might be a matter of total irrelevance to one individual, another could respond with considerable emotion. A devout Hindu, because of his world-view, would quite possibly be appalled. An uninformed animist might wonder if the purpose of these objects was somehow associated with magic or witchcraft. A mystic of another sort, could see it as little short of murder, and a hunter would be pleasantly interested if not envious.

Thus, we have to conclude that what we believe and the significance which we attach to any object or experience, cause us to react on a gut-level, not just a head-level. We really do assume so much without knowing why, or at least without pausing to think out logically why we do. Our *psyches* are constantly at work.

THE SLIDE FROM PSYCHICAL-INTUITIVE INTO PSYCHICAL-MYSTICAL

Of course psychical "knowing" can far surpass this intuitive, instinctive, gut-level reaction stage. An individual can deliberately give highest priority to internal subjective knowledge. It all depends on what a person believes. If it should be that one's world-view explains every occurrence as the working of unseen and unnatural persons or powers, beyond the grasp of the rational mind, it becomes totally "sensible" to react non-rationally. Consequently, the mystical side of psychical thinking takes over.

GOOD THINKING = RIGHT COMPONENTS, RIGHT STRESS, RIGHT ORDER

It appears that every normal person in any culture and in any age will have in them all three of these basic components or categories of thought. We all perceive things; we all have ideas about those things; we all experience subjective reactions. Our habits of mind and our conscious beliefs both cause us to spontaneously put special priority on one distinctive style of reasoning. We can shuffle the order around and put more stress on one than another.

If all the stress is placed on particulars, we get lost in the details without them having any real meaning or significance.

If our concepts are not applied to objective details, our world can become so abstract that we lose contact with the hard facts of reality.

When the psychical is given the preeminence, we lose contact with both concrete reality and sound logic.

Obviously, we should take all three — hard facts, ideas about those hard facts, and our feelings and preferences — into consideration. We need all three, but we most definitely cannot use all three equally, to come to a knowledge of the truth, and to be able to think straight.

Incidently, all three aspects of "knowing" can Scripturally be labeled as "heart knowledge." The heart is not subjective and intuitive only, standing in opposition to the intellectual "head knowledge." In Scriptural usage, it is a broad term which is used to include our rational natures, our wills, and our emotions. That which is of the heart is everything which is a part of man's cognition and awareness.[4] It is in contrast to the commonly-used subjectivist definition, which rejects the intellect, that we will use the term, "heart."

As stated earlier, the differences in ways of thinking come from the order of priority and amount of emphasis which is placed on each of the three components. How should they be put together?

In brief, perception just gathers information. It is incapable of evaluating and reasoning. To perceive is to observe, not to reason. Psychical impulses are by definition irrational and untrustworthy as a basis of sound thinking. They are feelings, not thinking. Conceptual reasoning alone is an intellectual process. It can compare specific facts with other information and with God-given principles and draw the necessary conclusions.

What process takes place in a person's mind when he consciously perceives something? As we observe and experience the events and objects in our environment, the automatic reaction is to determine how much significance this detail has to our lives. Is it very important, totally trivial, or somewhere in between? This evaluation may be made either conceptually or psychically — or, more likely, be a mix of the two. Which way predominates is mostly a matter of mental habit. By association, training, and personal preference, we have conditioned ourselves to respond either rationally or irrationally. Heads it's rational, tails it's irrational! No, I don't mean that mere chance is involved. Is our decision head-level or gut-level, fact or feeling? We have a choice, you know.

SUPERNATURALISM: RATIONALITY WITHOUT RATIONALISM

The early scientists, as Christians, thought their way out of the dark ages by learning to think conceptually. They looked for the principles and universal truths which explained reality, both natural and supernatural. They had no alternative way of thinking if they were to be consistent with the Biblical worldview which they had discovered.

Naturalism eventually came into existence because some of their successors decided that everything could potentially be reduced to the level of human reason. They left out the necessity of God revealing concepts which are beyond the limits of human reason. That's where rational*ism* came from. Naturalists continue to this day, to place the priority on conceptual thinking, but can't even properly justify it. The fact is that they've stolen our thinking style from us! Conceptual thinking is of value and worth stealing. They couldn't get along without appropriating it because, as we've already discovered, without that bit of inconsistency, their rationalism eventually leads them into irrationalism.

Christians have a God-given obligation to be rational but not to be rationalistic. There is nothing unspiritual or carnal about being logical and analytical. God is a God of order, and that includes orderly thinking. Indeed, spiritually-minded people will prefer a conceptual response over a psychical one. Both the concepts and the ability to think in concepts come from God. Supernaturalistic thinking is tied directly to ideas and concepts about reality which have been revealed to us by God. God has given us minds which really think and has put us in a universe which functions in an orderly way. He has said, in effect, "You find out why and how, but remember, there would be no orderly system if I had not organized it!" This makes sense, and without it nothing makes any sense.

THE PSYCHICAL PRIORITY ON THE IRRATIONAL

Psychical thinking places its priority on the acquisition of irrational "knowledge." The way in which the ideas are acquired can't be examined rationally and objectively, because they are internal, subjective, personal experiences which are beyond analysis.

Psychical thinking takes a strong grip when people develop the mental habit of acting on their intuition and impulses rather than on objective evidence. The justification is that one "just knows," is "positive" but can't say why, and has "a hunch," or is instinctually aware of things. We all do this when we unconsciously pick up clues, or put numerous little facts and impressions together. It isn't always necessary to sit down and work out a conclusion deliberately and consciously, step-by-step, but before drawing any firm conclusions, wisdom says to go back and check it out. It is, after all, possible to jump to false conclusions.

The change-over into the mystical thinking pattern takes place when we start crediting this inside knowledge to outside sources; for example, by assuming that the inspirations and attitudes which instinctively spring into our minds without conscious thought, are not our own, but are from God. In actual fact, *everyone* is familiar with the experience of having strong impressions or ideas leap into our minds, seemingly from nowhere. There are at least three theoretical possibilities for the source of these psychical notions: they are from God and are supernatural in origin; they are purely psychological and are natural in origin; they are mystical and are demonic in origin.

Which of these alternatives do we tend to assume is the source of our internal impressions? Much can be revealed as to whether an individual's personal bias is toward the conceptual or the mystical by his reaction to the statement in the last paragraph.

A negative and emotional response shows favoritism toward psychical, irrational thinking. On the other hand, to say, "I'll need to think this statement over carefully," is to give priority to conceptual thinking. How did you react? Which has a stronger hold on your mind, the force of the ideas, or the force of the emotional response?

Notes — Chapter Nine

1. Luke 1:1-4; John 20:30,31; Rom. 1:25; Heb. 11:6
2. For various ways in which these three aspects of cognition are organized, see:
    HESSELGRAVE, David J.: *Communicating Christ Crossculturally*; Grand Rapids: Zondervan Publishing House, 1978
3. Luke 10:22; I Cor. 2:6-13; Eph. 3:1-10
4. "Heart," as used in the Scriptures, is the innermost center of the natural condition of man; the center of the rational-spiritual nature of man; the center of determination, will, love, hatred, thought and conception, understanding, deliberation, anxiety, despair, fear, moral life; the place of origin and issue of all thought, words and deeds; the place of lusts, passion and conscience. See "Heart," page 462, in:
    UNGER, Merrill F.: *Unger's Bible Dictionary*; Chicago: Moody Press, 1957

10

Getting To The Heart Of The Matter

THE RELATIVE VALUE OF "HEAD KNOWLEDGE" AND "HEART KNOWLEDGE"

Were you ever afraid of having inadvertently given a false impression? If so, perhaps you'll be gentle when I say that you may have just received a false impression of what's being said. Almost certainly, many readers will be concluding that I'm saying that only intellectual knowledge is good and that emotions and intuitions are bad. This could be the unhappy and entirely incorrect conclusion drawn from the insistence that the church is in danger because supernatural and conceptual thinking are being lost in a shift back into psychical and mystical thinking.

This is *not* what we're saying! Both the intellectual and the psychical sides of our natures have an essential and proper role to play in our lives, and both will inevitably make an important contribution to our mental processes and spiritual growth. But both can be misused or neglected as well. What we are saying is that we must do our thinking with the head, not the *psyche*. The "heart" (as is commonly understood) is for feeling, not thinking. Feelings are good — good for those things which feelings were designed to be good for. The danger is in allowing psychical "knowing" to replace conceptual knowledge.

Conceptual thinking, not psychical "knowing," will allow us to understand reality as it is and to make sound, wise decisions. It makes it possible to not only know but also to obey God, to fulfill the commission which God has given man. Only by being certain of our understanding can we be certain that we are obedient.

However, God hasn't created us in such a manner that we can just go through the forms of obedience with cold and unfeeling logic and calculation. Man isn't built that way. Everything we do with our minds and bodies has a profound impact on our feelings, intuitions and emotions. Man can pretend to be a machine and deny his deepest instincts, but in the end it just won't work. If we pretend to be totally objective, factual, emotionless, we are, actually, artificial, unnatural, ungrateful and unholy.

SOUND IDEAS PRODUCE STRONG "HEARTS"

No, this is not a reversal of the ideas argued in the last two chapters. Priority in *order of arrangement* must be placed on supernaturalistic conceptual thinking, not on psychical, mystical non-thinking. Our knowing and our deciding must be rational and logical, in accordance with the reality of the supernatural realm, in harmony with the laws and principles of the cause-and-effect system of this natural universe in which we exist, and always in obedience to the objectively revealed will of God.

Ideas, especially true concepts about God, contain power! The awesome fact of the majesty of God, the vastness of His works, the grandeur of His purpose, His beautiful love and His dreadful wrath — how can any normal person fail to be overwhelmed and profoundly affected to the depths of his being by these? From the ideas which fill our minds, from faith and from obedience, the subjective side of man is reconstructed. Our experiences with the supernatural God reconstruct us spiritually, just as our experiences with nature reconstruct us psychologically.

"I CAN'T HELP BUT GET EXCITED!"

Before concluding that we are calling for a Faith which is head-knowledge only, emotionless, no "heart," and no intensity, remember that we have pointed out that every person has all three aspects of knowing as a part of his makeup. Man cannot deny his subjective nature, but he must not allow it to take over the function which the intellect was meant to have.

The ideas which we've been sharing up to this point have been full of emotional impact. Haven't we shared the excitement of reasoning together? Didn't it thrill your heart to think about the glories and wonders of God's Heaven? I don't know about you, but just picturing in my mind the creation of a brand new universe almost made me bounce up and down with praise and joy. Haven't we experienced here and there some sense of insecurity in regard to some of our long-held beliefs and assumptions? Haven't you felt concern, indignation, perhaps even — but I hope not — boredom? There is simply no justification for the supernaturalist being cold and unfeeling! Nothing could be more rational than for a Christian to develop a good devotional life — a life with time devoted to thinking about the glories of the Lord. The church is weak and anemic from insufficient spiritual nourishment. May the eyes of God's people be turned away from themselves and back toward the One for whom are all things and by whom are all things.

TRUTH AND ITS CONSEQUENCES

What is being stressed, though, is that the basis, the foundation, or motivation for our behavior must be rational, not psychical. Feelings are, by their very nature, non-reasonable, because reason stems from the intellect, not the *psyche*.

Impulses and intuitive reactions are always a result of something else. Emotional ups and downs — tension and peacefulness, feelings of fear or excitement, elation or despondency — whatever the mood — are never trustworthy guides because they could come from so many sources, either sound or silly. Solid facts or believable lies, common sense or poor judgment, changes in the barometric pressure, poor digestion, a glandular system either over- or under-functioning — anything and everything can influence the "gut-level" side of a person. An inevitable *consequence* of all that we do is the formation of conditioned feelings — subjective responses — within us. It shouldn't be objectionable to anyone to call them "feelings" because that is precisely what they are — non-conceptual, non-rational, non-objective and non-verbal feelings. (Although they are non-verbal, people certainly talk a lot about them!)

Intuition, impulses, and instincts are all signs or indications of preceding sources of influence. We are creatures of habit, both psychologically and physically. The psychical part of our nature is shaped and molded as a direct result of other influences. Consciously or otherwise, we learn by experience. By repetition we encourage within ourselves certain reactions and suppress others. Emotions are vented or restrained, attitudes are condoned or condemned. We are advised to trust or distrust the various impulses and instincts which we experience.

In addition to specific innate reactions, the form which an individual's entire thinking style takes is a matter of mental habit. The more, for example, that we make decisions or act on the basis of the way we feel subjectively, the more deeply ingrained is the pattern. Psychical-style thinking becomes a psychological habit. The next time around it will be all the more difficult to act rationally if our feelings tell us otherwise. Feelings are reactions — and reactions can't be trusted as grounds for making wise and sound judgments.

NEVER JUDGE BY HOW THINGS FEEL

Whether these psychical manifestations are helpful or are hindrances to us, depend upon whether they urge us to go in the right direction or not. The "truth" or "falseness" of our internal drives depends on the factual and conceptual accuracy of those influences and experiences which produced them. In and of themselves, they are proof of nothing.

"But I know what I've experienced!" we want to insist. Do we really know what we've experienced? Not necessarily. Until our subjective reactions have been measured, compared and tested by objective standards, the best we can say is that we know we've experienced something. Just being urged doesn't in itself make something right, but it's obviously much easier to act in a certain way when the power of the psychical side of our nature is urging and encouraging us on. Once our thinking is straightened out, we can afford to turn loose the powerhouse within us.

A USEFUL SERVANT OR A DESTRUCTIVE MASTER?

Feelings, like fires, are powerful, useful — and dangerous. Inner drives must serve us, not rule us. The flames within must be contained and controlled within the furnace walls of hard thinking and objective reality. The powers generated by those fires burning deep inside us become the driving forces which can move mountains and topple kingdoms. Restrained and directed they make man's work incredibly easier. But once outside the steel boundaries of carefully-constructed concepts and rational restraints, the fires rage as instruments of Hell itself.

11

What Do You Mean, "The Word Of God"?

God becomes extremely angry when someone claims to speak a "Thus saith the Lord" when it isn't the Lord doing the speaking! It's actually quite astounding to notice how casually the statement is thrown around: "The Lord really spoke to my heart the other day." Did He? Before claiming a revelation from God — because when God speaks that's exactly what it is — check out what He has to say about people who make such assertions when the ideas actually come from their own minds. It might be impossible for anyone to prove conclusively that I did or did not receive a subjective message from God, but *God* knows, and He demands that no such claim *ever* be untruthfully made. When God speaks, He leaves no room for speculation, imagination or innovation![1]

So there's disagreement on *subjective* revelation. That isn't so surprising. At least we're agreed on how God speaks to us through the *Bible,* right? Wrong!²

In spite of broad, general agreement that the Bible is definitely "God's Word" and "inspired," there isn't even proper agreement on the manner in which God speaks to us through that Book, our *objective* source of knowledge. Because of the influence which the mystical mind-set is having on the church, there are strongly conflicting attitudes toward the Bible and the manner in which it is to be used, which follow the same two distinct patterns which have been the subjects of our examination. There is, of course, a third attitude toward the Bible, the naturalistic view which treats the word anti-supernaturally and not as inspired by God,³ but that won't be considered now.

DISAGREEMENT OVER THE NATURE OF INSPIRATION

There are two distinctly separate concepts regarding what is meant by "inspiration." As we might have guessed, the supernaturalistic view makes inspiration fundamentally objective, residing in the uniqueness of the writings and the verbal messages which they contain. The mystical view, on the other hand, tends to assume that the reader also ought to be inspired to receive his or her own personal message.

DISAGREEMENT OVER METHODS OF INTERPRETATION

Directly related to the two conflicting ways of looking at inspiration, two completely distinct approaches have also been taken to the study and interpretation of the Book. One methodology follows the supernaturalist world-view and conceptual style of thinking. The other way of deciding what God is saying to us through the Bible is heavily reliant on irrational and mystical methods of "knowing."

JUST TO AVOID PROBLEMS: THREE WARNINGS

At this point it's important for us to clear up an essential issue, lest we have a serious misunderstanding. In the pages coming up are some touchy areas, so while continuing on *please* keep in mind that the issue is not the reality and work of the Holy Spirit! God promises the Holy Spirit will dwell in His people and perform His work. God always does His part. We don't need to fret over His reliability. He will keep His promises and fulfill His responsibilities whether or not we are aware of it. Our part is to be personally responsible in our use of the Scriptures. *Whether* the Holy Spirit reveals the mind and will of God is not at stake.[4] *How* the Holy Spirit intends for us to view and use the *Bible* is the area of conflict.

We also need to remember two warnings mentioned earlier: First, most of us aren't very consistent and will probably show characteristics which align us with both the supernaturalist and the mystical attitudes toward the Bible. Bible reading habits, some good and some not so good, have been developed and thus, molded our attitudes and study techniques. As a consequence, it's too easy to act on assumptions in regard to the Scriptures which we've never fully thought through, and to find it a temptation to work out our arguments to justify these preconceived notions. We're also often inclined to technically believe one thing but practice something quite different. That's why this chapter is important — and controversial. Let's think our way out of habitual reactions.

Secondly, remember that the issue is one of *priority*. Certainly, the use of the Bible defies a rigid "either/or" proposition; however a major change-over of emphasis is taking place in the church, and not for the good. The way of thinking which is attracting an ever-larger following is the mystical use of the Bible. No one can make a wise and intelligent choice between the two alternatives until the options and their implications are clearly laid out. Even if we decide that we need not choose one option to the total exclusion of the other, maintaining a balanced perspective means that we have to know what's being balanced one against another.

THE BIBLE: HOW DO WE HEAR GOD SPEAKING?

Now let's try to lay out, point-by-point, the two alternative attitudes toward the Bible in order to get at the nature of the differences. The question once again is, how does God use the Bible to speak to us? The alternatives being considered are the supernaturalist's objective approach and mysticism's subjective approach.[5]

INSPIRATION: **IN** THE WORD OR **THROUGH** THE WORD?

The supernaturalistic, objective understanding of inspiration, means that inspiration is *outside* the reader. God "breathed" His message into the Scriptures. Thus, the Bible is inspired, an objective, fixed standard against which man's ever-changing circumstances can be measured. It is a God-given yardstick which can be seen, handled, and transmitted from country-to-country and age-to-age.

By comparison, a mystical approach will define "inspiration" as residing as much in the *reader* as in the message read. To truly understand what the Bible is saying, one must be subjectively enlightened. We might say that God is speaking *through* the Word more than *in* the Word.

INSPIRATION: CONFIRMED BY OBJECTIVE OR SUBJECTIVE EVIDENCE?

How will the supernaturalist decide whether the Bible genuinely is from God or not? What method will he use to determine the actual "inspiration" of the documents? Objective methods, obviously. The truth and validity — or lack of it — of any document will have to be confirmed by critical examination of the subject matter, the manuscripts, and how they were transmitted, the outside evidence which might be available, and whatever other facts might have a bearing on the issue. The studies of "textual criticism," "apologetics" and "Christian evidences" depend upon objective evidence for the objective message in the Bible.[6]

"Confirmation" of God's message is also of vital importance to the person who uses a mystical approach. God speaks *through* the Bible, but also, he believes, through a wide variety of other means as well. Since no objective measurement can be applied to most of these, the ultimate test is subjective. External events may corroborate the message received, but "evidence" of the truth or falsehood of the Bible is given to us internally, intuitively and personally. Whereas the supernaturalist comes to a conclusion of truth, a conclusion of truth comes to the mystic.[7]

THE BIBLE: A SUFFICIENT OR INSUFFICIENT VERBAL REVELATION?

Because the supernaturalist places a very much higher priority on objective revelation from God than on subjective revelation, his conclusion is likely to be that the Bible is not only the genuinely and truly inspired message from God, but that it contains all the inspired message which man needs. It is from God. It is accurate. It is also sufficient. Man doesn't need some other kind of inspiration to know God and His will.[8]

Placing his priority on conceptual thinking, the supernaturalist finds fully adequate standards for faith and practice as a result of studying the character, nature and deeds of Jesus Christ, and the instructions and the principles revealed throughout the Bible. The details of every-day life are understandable in the light of *changeless principles*.[9]

"Plenary" inspiration, as this concept is called, is nonsense and unreasonable to the mystically-minded. The Bible is not sufficient. God speaks not only through the Bible but independent of it as well. Man needs this, for otherwise how can he make personal decisions or evaluate his experiences? He wants the direct leading of God in the many daily details about which the Bible never specifically speaks.

CHRIST-CENTERED OR MAN-CENTERED?

The first thing any good conceptual thinker will ask when he picks up a book is, "What's the bottom-line?" He wants to know the main theme, the basic idea of the entire volume. He knows that the details and portions of any book can only be properly understood in the light of the subject as a whole. The conclusion which he draws from applying this principle to the Bible is that it's all about God, as revealed by Jesus Christ. Jesus Christ — *He's* the "bottom-line." The message of the Bible is objective and the Object is Jesus Christ.[10]

It isn't easy, though, for fallen, selfish man to keep Christ foremost. How much more difficult it is to be truly Christ-centered when one's emphasis is not outward on the content of an objective message which centers on Christ, but is directed inward on the events and experiences located within a man himself.[11] It needs to be remembered that the various philosophies which are part of Humanism have made a mystical, irrational, blind leap into the subjective realm. To the humanist, ultimate meaning and purpose is found within the individual. "Christian" subjectivism also, by its very nature, is man-centered, although Christ may be given the credit. It encourages us to look inside for meaning.

DOES UNDERSTANDING COME BY STUDY OR ENLIGHTENMENT?

The Bible, being an objective message from God, can be studied by the same methods and rules which apply to other writings. Since it is a verbal communication, the inspiration rests in the words of the message. Therefore the exact words are extremely important. If improperly translated or transmitted, the message may be weakened or even lost. Words must be properly defined and points of grammar must be studied for their significance to correct understanding. The literary style of each section of the Bible — narrative, instructional, prophetic, poetry or prose — makes an interpretational difference which must be taken into consideration. Why was it written and to whom? Each part needs to be understood in the light of the over-all purpose and context of the document.

All of these facts, so significant to the supernaturalist, are trivial and of only minor concern to the subjectivist. By personal enlightenment and subjective revelation, the "Spiritually-minded" person will be taught of God. Doctrinal understanding must be revealed, often through the agency of the Bible, but always by the direct working of the Holy Spirit.

The differences between these two ways of learning are very closely related in principle to the next two aspects of how the Lord speaks through Scripture.

THE BIBLE: TO BE INTERPRETED HISTORICALLY OR IMMEDIATELY?

The supernaturalist approach looks at the Bible from an historical perspective. Here are sixty-six separate documents which were written over a period of nearly sixteen-hundred years, in three languages, on three continents, by over forty different men. Who was writing? Under what circumstances? What was the religious and political situation at the time? In light of the differences of culture, how would the people of the day in which it was written have understood it? God's character and nature can be better understood when we hear what He's said and see what He's done under all those different circumstances. The Bible can't be properly understood without taking historical settings into consideration.

The reason the supernaturalist is concerned about the historical circumstances is in order to identify the concepts and guiding principles which are above and behind each specific instance. No application can be made until the principle to be applied is nailed down. Which is an eternal concept and which a cultural application?[12]

As in the previous case — the distinction between the two ways of studying or learning — the mystically-minded person is not concerned with the historical, only the immediate. History is objective; his interests are not. The only concern is, how does this apply *right now* to *me*? A concept or principle as a universal truth hardly exists. In fact, a Scripture verse may be interpreted as meaning something entirely different from one individual and circumstance to the next, and may legitimately be applied completely out of its original context. The message, while channeled *through* the Bible, will be related very intimately to the present.

The popular "Typical" preaching style provides a convenient bridge for an easy transition from the supernaturalistic position to the mystical method of Biblical interpretation. The Bible is well-known for the numerous "types and shadows" which it contains. Types are rather like prophecies in events rather than words. For example, Adam was a type of Christ, Israel's crossing of the Red Sea foreshadowed Christian baptism, and the Old Testament tabernacle was full of symbolism. If the Bible states that an event, a person or an institution was typically symbolic, then it certainly was, but a popular trend is to treat any and every historical account as if it were a parable. The Bible is history and not just a book of mysterious symbolic stories.[13]

Have you ever been amazed at the hidden meanings which a preacher claimed to find in some common story? You had probably been under the impression that the account of Zaccheus climbing the sycamore tree was just what it says: a short man getting up where he could see. I'll bet you would never have guessed that this actually referred to his spiritual smallness, his effort to be a big-shot and elevate himself, and that the first thing he had to do to come to Jesus was to come down — humble himself. Surely only someone with special inside information can really understand what the Bible is talking about!

Not only does such so-called "typical" preaching lack Biblical authority (since the Scripture doesn't tell us that these hidden parallels are the true meaning of the events recorded there) but God's people are being misled into an entirely erroneous view of the Bible. God's book is not a book designed to *hide* information from us, but to *reveal* information to us.[14]

A UNIVERSAL MESSAGE OR A PERSONAL MESSAGE?

Viewing the Bible historically, causes the supernaturalist to recognize that it contains messages originally written to specific people who were in specific places, and who had specific needs. The entire Law of Moses was explained as being the legislative system for the twelve tribes of the nation of Israel after they entered the territory which God was giving them. Each of the Old Testament prophets delivered messages which only make sense when the events occurring in their day are taken into consideration. Paul wrote to Corinth to straighten out divisions and false teachings, discuss hair styles and head coverings, collect money to be sent to Jerusalem, and to pass along numerous personal greetings. He wrote elsewhere to request a co-worker to come before winter set in, for him to bring along a cloak and books which Paul needed, and to warn against a particularly dangerous opponent.

So how does it apply to us personally? Do these historical and personal references make the Bible any the less the word of God? To the contrary, they make God's will more understandable. Since the supernaturalist tends to automatically look at things conceptually, this isn't an insoluble problem. Biblical concepts are universal and timeless, and apply to every person.[15] The details make sense when he understands the concepts behind them. The reverse is also true. These historical references, the particular applications of those abiding principles, may perhaps be specific to the historical and cultural situations, but they are valuable and necessary examples of the application of the principles which the Bible teaches. They serve as illustrations so that we can understand the mind and will of God properly.[16]

The supernaturalist doesn't try to bind on everyone the command given to Moses to climb Mount Nebo and die there, nor Jesus' command to the Apostles to wait in the city of Jerusalem until power from on high is received. Those in themselves are not universal principles. Of course, "obedience to God" is a universal principle. Wherever there are direct commands to be universally obeyed, the context makes it plain enough.[17]

WHAT DO YOU MEAN, "THE WORD OF GOD"?

God is speaking to us in His verbally inspired Book. This profound fact ought to make an impact which is powerful, personal, and intimate, both intellectually and subjectively, on those who have ears to hear. Still, the principles are universal with applications suited to every conceivable situation.

The ultimate concepts to be learned are the character, nature, and deeds of God, as revealed through Jesus Christ, in order that we may saturate our minds with Him and become more and more like Him. Only by thinking conceptually, can we possibly understand the unchanging character of God and His will for all people. Rather than having a rigid legal system or a specific instruction for each little decision, the Lord's person must be honest and responsible enough to apply God's universal standards — His own character — to his or her own individual situations.

The essential nature of mystical-type thinking makes fixed, universal principles and concepts impossible, of little relevance, and entirely inadequate to meet man's personal needs. It is a personal message, not a general one, which the subjectivist seeks. What God has to say to him, he believes, won't necessarily be the same as what He says to you or to me. God has a personal plan for a man as well as a general one for all mankind.

The distinction between the universal and the personal outlooks is thrown into sharp contrast by the differing stress or emphasis which can be observed in various Bible studies and sermons. As the church slips further into mysticism, the parallel trend is to move away from in-depth study of the Bible, verse-by-verse or book-by-book. Expository preaching is almost a thing of the past. Instead, there might be lessons on topics of current interest using the "proof-text" method, or the sharing of personal insights, revelations, and experiences. The question is not so much, "What does the Bible say?" as "What does the Bible say *to me?*"

Perhaps the supernaturalist's attitude toward the study of the Bible seems intimidating to many people. Can the person of ordinary intellect ever hope to understand it? Certainly the ordinary person can, but not the *lazy* one! Firstly, the Bible can be understood because God is able to cause us to understand.[18] This isn't mystical, irrational insight we're talking about now, but supernatural intervention in the functioning of our brains, reasoning, and memories. When He opens the eyes of the blind, it is to see things as they are. The Lord can also be trusted to know when or whether to intervene.

In the second place, God fully intended for us to invest much effort into knowing His Word.[19] Serious study is necessary to properly understand the objective message from God. Good thinking is always hard work, and there are no short-cuts. To have the help of wise and knowledgeable teachers of the Scriptures is a great asset and can save us a lot of unnecessary time, but we will still need to always check out each idea with the written Word itself. Trust God. He's not only able to give us His message in the Bible, but to give it in an understandable manner.

THE POWER OF THE WORD: IN MESSAGE OR MAGIC?

The Gospel, we're told, is powerful.[20] To the supernaturalist, the power of the Bible is in the message which it contains. Ideas really do matter. What the idea is makes all the difference. It is, after all, the power contained in ideas — the re-discovered Biblical world-view with its various implications — which has changed the face of the earth so drastically. The supernaturalist depends on the Bible far more than most people realize. Logically he needs it, because his mind tells him that nothing in nature — including his mind — makes any sense or has any value unless it comes from a rational/supernatural origin, and he has no way to know what that might be apart from God telling him. Logic supplies the problem but revelation alone supplies the answer.

What is the Bible? Externally, only a book: paper with printing on the pages, bound together with cardboard or leather. However, in the mind of the supernaturalist, there is an awe-inspiring power which is to be found in the message of the Bible.

By comparison to this, there once was an old mystic called Johnny Appleseed who roamed the American frontier. It's said that he would leave pages torn from a tattered old Bible with the settlers whose cabins he passed. Later, if he came that way again, he would exchange that page for a different one. To him, it didn't matter if the page was read upside-down or backward. It was God's Word, and the power it contained would inevitably come through.

Mystics with somewhat more sophistication might use the Bible as a good-luck charm, a repeller of demons, a focus for the power of the Holy Spirit, a message-giver in place of the ouija-board and tarot cards, a crystal ball or a symbol of divine authority. As a divine object, it is seen as containing special use or significance. You don't know anyone like that? Think about it: did you ever know of anyone who expected to receive a prophetic message or divine directive by allowing the Bible to fall open and placing a finger at random on the page? What's the intrinsic difference? How about the person who uses his regular Daily Bible Reading as others, more openly paganistic, do their horoscope in the morning newspaper? Or the Christian lady who urges that a Bible should be left lying open because of the invisible power emanating from the pages?

From a mystical point-of-view, the Bible — the words, and

perhaps even the paper and ink, but not necessarily the original ideas of the passage — becomes a channel of power, a focus of force, an instrument through which God communicates with them as unique individuals. They have inside information, at least for themselves. While few would agree with old Johnny Appleseed, there is no shortage of those who expect the power of the Holy Spirit to use the Bible in new and surprising ways, with the words saying something totally different from the original, in-context idea. The power in the Word does not reside in the idea which finds its home in the intellect, the mind, but in the "Spiritual" event which takes place on an experiential — existential — level.

Summing up the differences, the comparison between the two attitudes is stark and shocking:

The Supernaturalist View:	**The Mystical View:**
1. The Bible itself is inspired.	1. The individual must also receive inspiration.
2. Objective evidence of inspiration can be studied.	2. Evidence is internal and cannot be objectively communicated.
3. The Bible is sufficient.	3. The Bible is insufficient.
4. The message of the Bible is Christ-centered.	4. The message of the Bible is man-centered.
5. Understanding requires disciplined study.	5. Understanding requires miraculous enlightenment.
6. Historical interpretation.	6. Immediate interpretation.
7. A universal message.	7. A personalized message.
8. Power in the ideas contained in the Bible.	8. Power in a mystical force working through the Bible.

AT THE CROSS-ROAD CONCERNING THE WORD OF GOD

Two different attitudes. Two ways of looking at and using the Bible. Yet we have only one Book. What shall we do with

God's written Word? Is it *equally* proper to use and gain information from the Bible in *both* of these ways? That is basically what's being done in practice, isn't it? It's amazing — and disturbing — to discover how we can claim to believe one thing while practicing something else.

Certainly God is capable of doing whatever He chooses. Has He made a choice? What ought to be our primary and habitual approach to the Bible? What expectations which we take for granted when we open the Book? The written Word of God, judged on its own merits, taking its own statements about itself, adding nothing — in short, studied objectively — forces us to conclude in favor of the supernaturalist outlook.

The Bible doesn't instruct us to understand it intuitively or irrationally, or for each reader to wait for his own personal revelation. The first choice, or the normal framework, ought to be the one which places Christ at the center, the one which treats the Bible message with the greatest respect, the one which is in harmony with The Message of the Bible Himself.

Once upon a time there was a man who got involved in a doctrinal argument with another Christian. Several times he firmly declared that he believed the Bible. Did he know, he was asked, that the Bible specifically stated what he was specifically denying? No, he said, he didn't know that. The open Bible was put in front of him. He gazed silently for a moment. "Well," he concluded, "I guess I don't believe the Bible as much as I thought I did!"

The moral? When it comes to the crunch, we might find ourselves between a rock and a hard place, between a belief we always thought that we held, and practices which just don't tally. We have thought that we were supernaturalists, objective and rational, holding to the time-honored verities of the Faith, standing solidly on the verbally-inspired, propositional, plenary

Word of God. And now some of us are beginning to wonder if perhaps there isn't a little truth in this other thing labeled "mysticism," simply because we're more like that ourselves than we had realized. Some serious re-thinking needs to be done.

I'm trembling as I look ahead to the dynamite topic in the next chapter!

Notes — Chapter Eleven

1. Deut. 18:20; Jer. 14:14-18, 23:16-38, 28:1-17; Ezek. 13:1-9, 22:28; Matt. 7:15-23; II Pet. 2:1
2. Mark 7:9-13; II Cor. 2:17, 4:2; II Tim. 2:15; II Pet. 3:16
3. I Thess. 2:13
4. Prov. 1:23; Isa. 11:2; John 16:13,14; I Cor. 2:9-16
5. The several questions which are discussed in the pages which follow may be studied in greater detail in such books as:

BOICE, J.M.: *The Foundation of Biblical Authority*; Grand Rapids: Zondervan Publishing House, 1978

GEISLER, Norman L., Ed.: *Inerrancy*; Grand Rapids: Zondervan Publishing House, 1980

LINSELL, Harold: *The Battle for the Bible*; Grand Rapids: Zondervan Publishing House, 1976

GAUSSEN, L.: *The Divine Inspiration of the Bible*; Grand Rapids: Kregel Publications, 1971

HARRIS, R. Laird: *Inspiration and Canonicity of the Bible*; Grand Rapids: Zondervan Publishing House, 1969

MICKELSEN, A. Berkeley: *Interpreting the Bible*; Grand Rapids: Wm. B. Eerdmans Publishing Company, 1963

PACHE, René: *The Inspiration and Authority of Scripture*; Chicago: Moody Press

6. Most Christian conservative scholars believe that there is indeed a "subjective response to the objective Word through the impetus of the Spirit." (R.C. Sproul, p.336, "The Internal Testimony of the Holy Spirit," in *Inerrancy*, cited above.) This is closely related to an individual's predisposition to believe, but cannot be adduced as "evidence."
7. I Tim. 2:4; II Tim. 2:25,26, 3:7
8. II Tim. 3:15-17; II Pet. 1:19
9. Psa. 119:9,11,24,105; John 5:39
10. John 5:39; Acts 10:43, 18:28; I Cor. 1:22-24, 2:2, 15:3
11. Prov. 3:1-5; Jer. 17:9; Matt. 16:24,25; Gal. 2:20

12. A good treatment of this hermeneutical problem is found in:
    KAISER, Walter C., Jr.: "Legitimate Hermeneutics," in *Inerrancy*, cited above.
13. For example, see:
    FAIRBAIRN, Patrick: *The Typology of Scripture*; Grand Rapids: Baker Book House, 1975
    HABERSHON, Ada R.: *The Study of Types*; Grand Rapids: Kregel Publications, 1957
14. The problem which the "natural man" in I Cor. 2:9-16 has, is not in technically being unable to understand the revealed message, but in "accepting" it — believing what is "foolishness" to his world-view.
15. Rom. 16:25,26; II Pet. 1:20,21
16. Note the New Testament use of Old Testament characters and events. The book of Hebrews has numerous examples. See also I Cor. 10:6,11; Heb. 4:11; II Pet. 2:15,16
17. E.G. Matt. 28:18-20; Gal. 5:1-26; Col. 3:1-17; II Tim. 2:2
18. Sproul, cited above, points out that "illumination," in Reformation theology, is never independent, but is always associated with the study of the Scriptures. Illumination is enlightenment which produces an accurate understanding of the Word.
19. Acts 17:11; I Pet. 1:10-12
20. Rom. 1:16; I Cor. 1:18

12

Who's Responsible?

"Not my will, but Thine be done." A significant statement, these words of Jesus, and an especially important example for anyone who honestly cares about being obedient to the leading and purposes of God. To know and do God's will is the basis of all discipleship, of all aspects of the Christian life.

But *how we know and recognize God's will* is not at all self-evident. The "will of God," in this context is referring to God's intent, His prescribed outline of events. Jesus prayed, "Thy will be done" because He desired to be precisely obedient to His Father's plan and purpose, in regard to His impending death.

Actually the big question of God's plan in the world can be subdivided into at least three more specific questions:

First, to what extent does God have a great universal and foreordained plan which explains the events which go on around us? Second, does God have a unique and personal plan worked out for each individual Christian's life? And third, how does He reveal His plan, whatever it might be, so that we can recognize it? We must know whether God has a plan for this world and for our own individual lives, and we must know how to recognize that plan, lest we find ourselves going against the will and purposes of God.

The answers which one gives to these questions will be directly related to that individual's belief-system, his conclusions about what does and doesn't exist, and why things happen as they do. Because of the differences which exist between the supernaturalist and the mystical world-views, the problem of discerning God's will is in many ways a "watershed," a dividing-point which causes the two streams of thought to flow farther apart with every decision made.

Here, as in almost every other area of Christian thought and practice, the very existence of two incompatible ways of thinking is largely overlooked. This oversight has sad consequences and is a cause of perplexity and trauma for many earnest Christians. Failure to distinguish between the supernaturalist solution and the mystical solution to learning God's will for our lives, inevitably results in guilt-feelings, confusion, and inconsistency as the two very different sets of assumptions become mixed up and jumbled together.

WELL, DOES HE OR DOESN'T HE?

The mystical position leans very strongly toward the belief that a fully-comprehensive plan of God does truly exist. This plan is viewed as incorporating not only a general and universal program, but also a specific blueprint for the life of each individual Christian. Purpose is seen in everything. God must and does, they believe, reveal the details of His plan for His people step-by-step, showing the particular path He wishes to be taken. In comparison to this, the supernaturalist's position is that God cannot be expected to reveal such a complete and detailed plan because He doesn't necessarily have one. God leaves considerable latitude for man to act on his own initiative and still remain pleasing to his Lord.

If the mystical approach to knowing God's will is the correct one, the supernaturalist will find himself insensitive to God's purposes and living outside of His will. If we decide to act on personal initiative, might we not be guilty of "running ahead of God," and rejecting *His* plan for our present and future activities? Rather than run the risk of getting outside of His will for our individual lives, shouldn't we be "waiting on God"?

If the supernaturalistic beliefs are true, the mystic will be the one who finds himself in danger of opposing God. If the Lord does intend for His people to use personal initiative, the mystic will be guilty of constantly *sinning by omission*. By waiting for God to specifically reveal each step before it's taken, we may well find ourselves neglecting the obvious opportunities which present themselves, and allow the world to grab the initiative away from us. On the other side of the same coin, the mystic might also find himself attributing to God many events and circumstances which were neither His will nor doing.

Shouldn't the church be the first to take advantage of the latest technology, be the most alert at recognizing trends, and be the strongest at countering threats? Where is our foresight, our planned invasions of enemy territory, our program to cause trends as well as counter them? While the sons of this age are successfully manipulating the mind-set of multitudes of people, we must not be fighting yesterday's battles. To the supernaturalist, the opportunity or need alone, may be sufficient calling. To the mystic, God must reveal His specific desire regarding whether, when, who and how.

Only after we've clarified our world-view can we be assured that we are staying within His will. The alternative to settling the issue and gaining this assurance, is to remain in doubt concerning God's desires, and to be guilt-plagued for fear of neglecting our spiritual duty. We can never make a decision on any spiritual matter and have peace of mind about it, unless we are certain that we haven't infringed on God's will. Confidence concerning God's wishes is no trivial matter.

As we analyze the issues, the importance of restoring and retaining the supernaturalistic world-view should become very evident. One of the major effects of the slide into mysticism is that it can seriously mislead our efforts to know and act on God's will for our lives. The first step of this analysis is to study the question of God's universal plan for the world.

A CIRCUMSTANCE OF NATURE OR AN "ACT OF GOD"?

How many times have we heard a Christian bravely accept a tragedy, for example the loss of a loved one through accidental circumstances, with, "God's will be done"? Is it true that God is personally responsible for the flow of events which make up the history of the world, both human and natural? To what degree is God directly instrumental in the multitude of incidents which occur around us? If God is not responsible for everything of significance which occurs, then are the good occurrences from God and the bad ones from Satan? Could it just possibly have been chance, or accident, or human carelessness or irresponsibility, or the deliberate act of some other person?

The issue in this first question is whether or not, behind each and every experience or event, there is either a supernatural or a mystical cause; some explanation beyond the natural, or some invisible personage who is responsible. The idea of *chance*, especially to the mystical mind, seems so — worldly? At least some find comfort in believing that there is a hidden, divine purpose behind the apparently senseless events which we all experience. Satan is more likely to be blamed only for those problems which are less than fatal. In any case, a trend in the church, as it moves farther into mysticism, is to attribute everything that occurs to the activities of two competing powers, the good and the evil, represented in the persons of God and Satan. Nature doesn't enter into the explanatory picture very much, if at all.

DON'T ALLOW MYSTICISM TO WEAKEN BELIEF IN GOD'S SOVEREIGNTY!

God is not limited, however! He is not bound to be the cause of each and every event which happens. He is free to work His sovereign will without this negating His own creative acts in nature. God's will does not necessitate that every event be determined and fixed by God. He can actively intervene in nature — or not do so — as He wills. A form of Divine determinism is no more justifiable or true than is naturalistic determinism.

God is truly sovereign, truly transcendent, truly independent and superior to the natural universe which He's created. It is untrue to say that there is no distinction between the "laws" of nature and the deeds of God. "Mother Nature" is not another name for God! God is supremely Personal. He thinks and wills with deliberation and acts with the ultimate in purposeful intention. By His sovereign will, He has chosen to set up a system, this natural universe, which will carry itself on without the necessity of His constant agency in initiating every event. Nature, in total contrast to its Creator, is totally impersonal and does nothing "on purpose." The laws of nature simply *are* —

they are our observations and descriptions of the mindless functioning of the system, and we call them "laws" because of the regularity and predictability which we see. God, in His sovereign capacity, chose for this to be so. In our thinking, we must allow God to be supernatural, and nature to be natural.

DON'T LET MYSTICISM NULLIFY GOD'S CREATIVE ACTS!

God hasn't created an organized universe to nullify His own act by overriding or ignoring the system of natural cause-and-effect through His own continuous creative activity. Nature is real, and the basic structure of events is the outworking of natural forces.

God, in His infinite wisdom, created this wonderful universe as a system, orderly and self-perpetuating, designed to last over a period of time sufficient to accomplish His will. As and when He wishes or deems necessary, God reaches into this system from His supernatural realm to intervene. But this doesn't destroy the fact that it remains a natural and "impersonal" system, any more than when the builder of a complex machine reaches into it to adjust, alter, or repair his handiwork. The normal framework continues to be that of an impersonal, mechanistic-like *thing*. If there were no agents of change, naturalistic determinism would indeed be true, because natural functions would explain everything. As it is, acts of nature explain only a portion, but a real portion: the basic structure of the events which get changed by the agents.

The mystical assumption (into which the church is sliding) tends to see God — or Satan — behind every significant event, act or circumstance. Absolutely nothing is believed to happen by chance, or just by the nature of things. Christian experience does not include coincidence. As is the case with more overt pantheism or animism, there is behind everything which occurs, some kind of invisible power or force.

Don't negate God's creative work! He said it was good, and it is. Creation has meaning and significance. It provides an orderly pattern composed of space and objects and time within which all of God's creatures, Christians included, are able to live and function. This natural framework isn't an illusion. We are not being tricked by the appearance of natural law and order while in fact each event is the specific manipulation of either God or an evil counterpart named Satan.

Nature has meaning only if we allow it to be truly natural. Does God cause each individual flower to blossom? Is the plumage on every beautiful bird a product of the direct action of the Creator? Why does a particular human baby come into the world? What causes the storm, the disease, the unforeseeable event, whether beneficial or tragic? Is God the direct Agent in each and every instance? Pressing the concept to its ultimate, we must decide such basic issues as whether it is true or untrue that God completed His creative work after the sixth day, and whether or not it is true that new living creatures are the result of God's plan by which each form of life naturally and spontaneously reproduces after its kind.[1] If the "machine" won't run by itself, it isn't yet complete!

Notes — Chapter Twelve

1. Gen. 1:11,12,21,24,25,28, 2:1-3

13

Man — An Active Ingredient

A few hundred years ago some Christian men took a long, hard look at the Bible, then looked around at nature with a new world-view. They saw that nature was truly natural. It was not blasphemous to treat things as *things*. We owe the origins of modern science to the thinking of people who took God's creation seriously.

But how can God have a plan, with a guaranteed outcome, unless He controls whatever happens in the meantime? How this can be isn't all that difficult to illustrate. Picture a novice playing chess with a grand master of the game. The novice may freely and legally make any move he chooses, within the possibilities and limitations which the rules have long ago established. The chess master can even guide and instruct the novice, allow him the broadest of latitude, give honest advice —

and still be capable of winning. The superiority of God's knowledge and ability over all other minds collected together is vastly greater than the gap between the greatest champion of the game, sitting across the chess board from the rawest neophyte. Part of God's "game-plan," especially the opening moves and the end-game, He has done and will do Himself. Part of His plan God leaves in our hands.

God isn't playing games however. He is not a loser and He doesn't want us to be losers either. It is His universal will for us that we choose to be on His side. Once we have made that choice, God tells us what He wants done and what the final outcome will be. Our task is to work toward that end. He's laid down the limits within which we must live and act: the structure of our universe, the limitations of our human natures, the ethical standards to which we are to conform, the assistance and guidance which He offers and the specific goals toward which we are aiming. Within these boundaries it is God's will that we think, learn, make decisions, use our own initiative, and move toward an ending in which both we and God are winners.

DON'T ALLOW MYSTICISM TO DESTROY MAN!

Man is *Man*! He isn't only an animal, a nonentity. God made man in His own image, a rational being with a mind and a will that really work. Man is more than a spectator to God's great plan, more than a pawn, more than a bystander, innocent or otherwise, helplessly caught in a titanic struggle between two vastly powerful and inscrutable beings. He is more than an inert tool. Man is also an *agent*!

Man has genuine significance because he is an agent of change. He makes decisions which are real decisions. History is different because of the choices which man makes. Great natural disasters, such as floods and earthquakes, and the many other circumstances which are beyond man's ability to control, are often called "acts of God." Medieval and pre-technical man thought this way. Anything man didn't understand or couldn't control must be an "act of God"! But if this were literally true, man should never try to control *anything* because he would be opposing God. Did you ever stop to think that when you water your lawn or pull up a weed, you're interfering with the natural course of events? Is a dry lawn and growing weed an "Act of God"?

God Himself has commanded man to take control over nature. He is to be a co-agent with God. Because man has fallen so far from what God intended him to be, nature more nearly controls him than he does nature, and yet, some things have been accomplished, especially since the restoration of a Biblical world-view and the consequential development of modern science. For example, through the Middle Ages the population of Europe didn't increase, primarily because starvation kept it static. The population explosion is largely the result of man's increased control over drought, disease and other natural disasters. On the other hand, man, through war, is also responsible for producing some of the worst disasters in human history.

Man is most certainly significant and capable of altering the course of history, and he can alter events for the good as well as for the bad. To be an active agent and deliberately cause things to happen is not in and of itself wrong. Our decisions are not automatically sinful and contrary to the will of God, simply because we are the ones who do the deciding. Otherwise, the only decision we could make which God would accept would be that of passively making no decisions. To refuse to make decisions is, in itself, a decisive act of omission which has an influence on the flow of events.

Man is by nature a decision-maker. Christians are instructed to pray for wisdom in order to be good decision-makers.[1] However, if man is not a legitimate agent of change, and either God or Satan are the actual decision-makers behind every significant event, human decisions must be impossible. If God has a comprehensive plan already worked out, human decisions must be sinful. Neither is true.

COMMISSION, NOT COMPULSION

The same Bible which reveals the supernaturalist's worldview, the belief-system which agrees with both nature and man's character, tells us that God certainly does have a purpose in mind for man and the world. The Book, like the thriller which it is, tells of the unfolding battle throughout history for man to fulfill that purpose. Satan and his mystical forces are out to thwart man's efforts to obey God. We can read the end of the book to know that it has a happy ending. The good guys are on the winning side.

However we need to understand that the end doesn't come with man, by his own efforts, conquering the enemy. After things have gone on long enough, God will step in and put a stop to hostilities and an end to enemy resistance. It is the ultimate victory, not all the intermediate battles, for which God has determined the outcome.

All of this is simply to say that God has *commissioned* rather than *determined* that His will be done. God will do His part. As far as man is concerned, however, whether God's will is done is a matter of obedience, with no appearance of divine compulsion. It involves that which we *ought* to do rather than what we are *compelled* to do. Words like "obedience," and "ought" don't even make sense if man has no alternative, no choice in the matter.[2] On the other hand, we truly don't have any alternative about conforming to the "laws" of nature. While we're in the process of carrying out God's orders, we must take into consideration and take advantage of the functioning of the natural universe in which we exist. God doesn't exempt us from the natural cause-and-effect, and we must expect the same impersonal treatment from the system as anyone else. We can ask Him to intervene of course, but we have no right to ask Him to destroy the system! Neither does He exempt us from the necessary task of making difficult and honest decisions. Being an agent of change — and to be anything less is to be less than man — carries with it an awesome responsibility.

The Bible reveals God's universal will, His over-all plan which applies to all of His people. His will for the church and for each Christian is expressed in the form of instructions to be obeyed; it is not simply an announcement of some fixed and inexorable series of events. In fact, more than giving us orders regarding what to *do*, He tells us what to *be!*[3]

It probably isn't difficult at all for the average Christian to accept that God intends for His church to make an impact on the world, and for mankind as a whole to exercise some measure of control over our environment. God's universal will isn't much of a problem. In fact it shouldn't be a problem at all. Of course God has a universal will, and He works that will through His church and through mankind in general. The crunch comes over whether God's will is determined or commissioned. The trauma arises over the insistence that God wants man to make decisions, to take initiative and have plans and goals which he himself has made.

The influence of mysticism shows itself right here. It is mystical thinking which says that the universe is permeated with a power or consciousness with which we must harmonize. Mysticism doesn't like personal initiative very much. It is Biblical supernaturalism which sees man as an agent of change, acting and not only acted upon.

Mankind bears a specific responsibility in the realization of God's universal will. It might be discovered that the idea behind the popular catch-phrase, "being used" is more compatible with the mystical "holistic," "ecological" philosophy of an all-pervading Life-force than it is with the supernaturalist's belief in God's commission to man, to become an active agent of change in the natural world.

BUT WHAT ABOUT ME?

Let's move, now, from the universal category, mankind as a whole, to the most specific — that's each one of us — and whether or not God does in fact have a unique and individual plan worked out for the life of every Christian. It's easiest to say "Yes, of course," but we'd better check out our position carefully. Remember, it's knowing and doing God's will which is at stake, a matter too important to base on ungrounded assumptions.

The very fact that an individual can ask this question explains why God's plan is not a fixed and pre-determined set of events, nor a series of moves by a God who manipulates or causes each and every event to occur. "Man," collectively speaking, is a bunch of individuals. You are self-conscious enough to wonder about your own individual place in the universal plan of God. You're not like a bee, which is locked into the bee-system, nor like the spawning salmon which is driven by unreasoning instinct to swim to its place of birth.

It must be obvious that what applies to the *whole* applies to the *parts* as well. What is true regarding God's universal will for mankind in general is also true of each individual. God requires the whole — all of us collectively — to use initiative, because He

has given this responsibility to each of us as individuals:
—It is the individual who possesses a mind which thinks real thoughts and makes real decisions.
—Each individual is an agent of change.
—Every man's personal experiences are lived within the framework of God's natural cause-and-effect system, from which no one is exempted.
—Each one of us must take personally the commission which God has given, shouldering the responsibility to make decisions and live a life which is conformed to the pattern He's revealed.

Because we are separate individuals, each with the capacity to think real thought and make real decisions, and because we're not the same in situation, ability, maturity or opportunity, the Lord expects each of His people to honor his commitment. Each member of the body fulfills its own function in its own place.[4] The example of Jesus Christ, His character and the principles revealed as being His universal will — form the concepts which indicate what God would have the individual do and be in each particular situation.

The point of all of our thinking and reasoning in this chapter has been to demonstrate the fact that God has given a measure of true *freedom* to His human creation, both individually and collectively. His plan is not so detailed and comprehensive that every event is the result of a divinely determined blueprint. To be frank, I have a serious suspicion that there isn't anyone who consistently lives as if each and every event that occurs in his life is the direct result of either Divine or demonic powers, although the claim is not uncommon. A thousand conscious and unconscious choices are made by each person every day. Although it may seem more "spiritual" to deny coincidence and the influence of impersonal natural causation on one's life, we all know it's there just the same.

Isn't God wonderful? Every one of us ought to praise Him daily for being made in His image, an agent of change, capable of thinking real thoughts and making real decisions which have a real influence on the eternal outcome of events. You — "you" being in the singular — are significant in God's scheme of things. You are important as a free-functioning person, not just as a cog in a machine. You have a commission. God's great and eternal will includes you!

Notes — Chapter Thirteen

1. James 1:5
2. Acts 5:29; Rom. 6:17; Gal. 6:2; I Thess. 4:1
3. Rom. 6:6, 7:4; Eph. 4:12-16
4. Rom. 12:3-8; I Cor. 12:12-27; Eph. 4:15,16

14

But I Need Help!

The third part of the big question — the nature and extent of God's plan — is this: how does God reveal His will to us? Can we know for sure if God has something specific and personal in mind for us? How can we be certain of God's leading?

The task of executing the broad and universal plans of God involves personal initiative, but does our Lord leave *all* of the details, the methods and procedures up to our own individual discretion? Among the multitude of situations and decisions with which each individual must deal, surely God *sometimes* has a special mission in mind for an individual, some task to be performed or niche to fill. Yes, surely He does, but to say this doesn't necessitate that He must therefore have a complete and comprehensive outline prescribed for our individual lives. It doesn't require that we stop being genuine agents of change in order for Him to be able to intervene as He sees fit. Just as nature is an "open system," so are our lives.

How then am I to know when God has something special in mind for my life? On the other hand, how can I tell when He wants me to take the initiative? Will my personal goals and plans conflict with His special intervention in my life? How can I avoid the error of confusing my own choices and decisions with those of God?

Sometimes the decisions are so big, and our sense of personal inadequacy so great, that we need all the assurance of God's blessing that we can get. When my family and I arrived on the mission field many years ago, a firm belief in the leading of God was vitally important to us, especially when times were difficult and discouraging. Reminding ourselves that we were "in the center of His will" for our lives, was all that kept us going at times. Yet He hadn't sent us a telegram, and no verse of Scripture said, "Stu Cook and family: go to Africa!" Subjectively we had felt "moved" to go — but also, in some ways, felt moved to not go! We prayed, we discussed the possibilities with the leaders of our church, but the final decision had been on my own shoulders. It was God's will!

Now, years later, we're as convinced as we were then, that it was not only our decision, but God's will for our lives. However, looking back objectively at that decision, there was always at least the possibility that it was His desire that we go *from* the ministry we were then in, and that it was up to our own discretion — with God's blessing — as to the specific ministry we went *to*. The Cook family and their Heavenly Father were in agreement, and simply because the move was God's will didn't necessarily indicate that He had an iron-clad plan exclusively for us. He was *willing* for us to go. His blessing was on the decision, because it had been prayerfully made in the light of God's universal will as it applied to our particular circumstances.

When people ask me why we went to Africa, there's no simple answer. What went on down deep inside me was not proof that God had specifically foreordained that move. Circumstances — "confirmation" in the fortuitous turn of events — might have been purely coincidental. God has never promised to answer prayers with subtle signs and omens. In fact, there's no avoiding the necessity of making decisions, of being what God made us — agents of change.

HOW DOES GOD SPEAK?

Christians ought to remember that *God is under no obligation to personally and specifically give directions to individuals if He doesn't choose to do it.* He wants us to make plans and decisions by our own initiative as well. In order to grow up in every aspect to be like Christ, and to develop a character like His, it is important that we exercise our judgments, use our wills and act with discipline. God desires for us to be so mature in the Faith that there is no clash between His goals and attitudes and ours.[1]

Sometimes circumstances force us to make important decisions quickly. A brief prayer for wisdom is all the time allows. To make no decision is a decision in itself. On other occasions, one is able to spend considerable time in prayer, study, perhaps research and meditate over an issue, as well as to seek out sound advice and consider the pros and cons. Why not? God wants us to be fully aware that we are agents of change and have a degree of responsibility for the direction in which events go. In fact, we are responsible whether we believe it or not.

Secondly, *we don't have to know whether or not it was God Who put an idea into our minds, or manipulated circumstances.* It isn't necessary for us to know whether He does or does not have a particular detailed plan for our lives. Why is such knowledge unnecessary? One reason is because God is *God*, and is perfectly capable of working out His will for us or in us without telling us. We must give account to Him, but He need not give account to us.² In addition, never has God indicated that we need to know, nor has He promised to keep us updated on His actions. It isn't even necessary on our part to justify whether or not He has a complete and comprehensive plan for our lives — an idea that's difficult, if not impossible, to justify Scripturally.

True submissiveness doesn't demand of God that He tell us "Why." Trying to "second-guess" God is neither a wise nor a respectful practice.³ There is no guarantee that we'll ever know whether some particular event was an act of God or whether that's simply the way things happened to turn out. It doesn't particularly matter anyway. Our job is to get on with business, the carrying out of God's commission, His universal will as it applies to our particular situation. God can always be trusted to do His part, which frees us to meet our own responsibilities. He doesn't need us to supervise Him! Faith trusts God; arrogance and doubt say, "Show me!"

Thirdly, *God still speaks through His Word*, and we'd better not overlook that fact in our search for direction in personal and contemporary situations. The sufficiency of the Scriptures is still at stake. When we do not misuse it mystically or irrationally, the Scripture is a totally adequate source of information for the motives and aims which we ought to have, the moral standards, attitudes and values which must be ours. Remember, we must not only do the right things, but do them in the right way, as well. In addition the Bible gives many clear principles and direct commands to be applied in every situation.

God leads us by means of the Bible in another extremely

valuable way. In fact, this is, without doubt, the most significant and important point of all. The purpose of the Scriptures, in the final analysis, is to show the character, nature and deeds of God as these are revealed by Jesus Christ. The more we have saturated ourselves with Jesus Christ, the more we find ourselves spontaneously reproducing His way of thinking and acting in our own lives. His attitudes, mannerisms, approach, and style inevitably take over and dominate our thinking. It is the way God constructed us: what comes out is the result of what went in. The spiritually-minded man, the man who has consciously and deliberately set his mind on Jesus Christ, finds his thinking and his will in ever closer harmony with God's. Learning how to *be Jesus Christ* in our daily lives, learning to act and think as living extensions of the life and ministry of Jesus in our own situations; this is true spiritual guidance.

When we were little children, we needed constant supervision, discipline, and detailed instructions about every little thing. As mature adults, with characters which have been developed by the training received as children, we find that it is just in our natures to respond in certain ways. If someone asks us, "Why did you do that?" we don't have to answer, "Because my daddy told me to act like that!" In fact, we'd probably have to stop and think about our answer, since it came from our character, not any longer from a direct order.[4]

It is an immature faith which finds itself unable to make difficult or unpleasant decisions, and doesn't know what is in harmony with the character of Christ. May the day come when we will, unbidden, weep over Jerusalem if He would have wept, or cleanse the Temple if He would be cleansing it!

Fourthly, *God leads us through the Holy Spirit* Who works in each Christian. How do we know we have the Spirit? Not because of subjective feelings or internal, non-rational experiences, nor by signs and wonders, but because He always keeps His promises.[5] How does the Spirit lead? Obviously, we can expect that leading to be supernatural in nature rather than fitting the mystical pattern. Of course, one way He leads, is through the Word, the sword of the Spirit. There's no promise that we'll have personal direct revelation — built-in "telegrams" from God. Notice the prayers of the Apostle Paul which can be found in nearly all of his letters.[6] They can be summed up as being appeals to the Lord for His people to be able to clearly know the character, nature and deeds of Jesus Christ. That is precisely the ministry of the Holy Spirit — to reveal Christ.

We'll talk more about the Holy Spirit later. This is an important issue, because the work of the Spirit is all too often viewed mystically rather than supernaturally. There is nothing irrational about the Holy Spirit, nor about the way the Spirit leads. God's Spirit doesn't specialize in intuition, urges or gut-level reactions. Because of the working of the Spirit of God in our lives, we are able to know and become more like Christ, so that we don't need to be afraid because God chooses to allow us to make decisions of our own.

One observation about the majority of God's direction in our lives which seems significant: most of it might be called "second-level" leading. It appears that He is more likely to directly intervene by making alterations in the decision-maker than in making the decision!

THE MYSTICAL ALTERNATIVE

The influence of mystical thinking can be clearly recognized in various methods which are popular in the search for the leading of God:

There is the "Open Door" method: we can pray and then take the course of least resistance. The problem is that the course of least resistance always tends to go downhill. The most readily-available opportunity isn't automatically the right choice, just because we prayed first. Don't expect God to answer a prayer which is contrary to His will,[7] and it is not His will that we avoid making careful decisions. Some open doors lead to disaster. Some closed doors need to be battered down.

Then there is the "Fleece" approach. Ask God for a sign. What did Jesus say about people who ask for signs? It is surely a "wicked and perverse generation" which makes a habit out of doing this.[8] The fleece approach comes from the story about the Old Testament hero, Gideon.[9] God certainly had a special task for him. But it isn't possible to consistently use the unique leaders of the Bible as examples of the way God universally works among His people.[10] Zacharias, the father of John the Baptist, also wanted a sign to remove His doubt — and was struck deaf and dumb.[11]

Closely related to the fleece, are the mystical abuses of the Bible such as are mentioned in Chapter Twelve. There's the "first-scripture-into-your-mind" method. And the "let-the-Bible-fall-open-and-stab-your-finger" method. And the "draw-a-card-from-the-promise-box" method. And the "see-if-you-can-make-the-daily-Scripture-reading-fit" method. God never promises to honor such non-rational non-methods.

Another shift into mysticism is to listen for an inner moving of the Spirit; subjective urges and instincts which incline us toward a particular course of action. Everyone has hunches, and they cause us to pay closer attention to some detail than we might have otherwise, but they are not safe sign-posts to the Lord's will. It's difficult to believe that a wise person would make a firm decision based only on subjective inclinations, being convinced that it is God's clear instruction, and then doggedly follow that course of action through thick and thin.

Rather than showing the clear leading of God, to obey inner movings will result in instability and erratic behavior.

As with all totally subjective movings, whether literally from the Lord or not, this is not God's method of communicating His plans. The supernaturalist will be aware that he is a fallen creature, and he knows better than to trust the complex, inconsistent, and untestable workings of his inner man.

The moving of the Spirit within: consider a common statement such as, "The Lord really spoke to my heart this morning!" "Heart" is probably the key word here. I don't recall ever hearing anyone say, "The Lord really spoke to my head...." The intention behind such a statement is probably not really to claim to have received an actual objective message, a "Thus-saith-the-Lord" verbal-type of revelation. More often than not this is a nice — but misleading — cliche, a way of saying that one has been impressed with a significant thought, realized the great importance of some spiritual concept, or has felt a strong urge to do some particular thing. Was it the Lord who directly and miraculously intervened in that person's thinking? Perhaps, but perhaps not. If our thoughts, activities and conversations are full of the things of the Lord, our subconscious minds ought to be running over with Christ-centered ideas! Non-Christians also have flashes of insight, hunches, strong urges within, which come from the things which occupy their minds.

What about prophetic utterances?[12] How do we know whether a purported prophecy is indeed supernatural instructions from God or a psychical sidetrack? A prophet of God is a person who speaks for God. A prophetic utterance is a direct message from God which has been received and passed on by the prophet. The message is, in every sense of the term, divinely inspired, as binding as the Scriptures. Scripture cannot be broken; neither can a prophecy fail — if it is indeed from

God.[13] Many false prophets have gone out into the world. There are also Scriptural references to prophets who really did have a message — but given by evil spirits.[14] To falsely claim to speak with a direct message from God, or to accept such a false message, is a terrible thing to do. It's an extremely serious declaration to say, "I have a message from God."[15]

Do prophetic utterances from God still occur in the church today? Some people to whom you speak, will rigidly deny the possibility that there are ever prophetic utterances in this age. Others go to an opposite extreme by accepting not only that it can and does happen, but tend to give the benefit of the doubt to almost any statement which professes to be from God. Rather than make dogmatic statements or react against extremes and abuses, let's look at some Scriptural principles applicable to this question. The supernaturalist way of thinking is to analyze details such as this one in the light of the Biblical concepts which apply.

First, is the message clear and unambiguous? When God speaks, there's no question about either the fact that He did it, or the clarity of the message given. When God speaks, it's loud and clear![16]

Secondly, is the message in total harmony with the contents of the Bible? The Holy Spirit never contradicts Himself.[17]

Third, does it say the same thing as the Bible has already said? If so, God would be unnecessarily repeating Himself, and either the "prophecy" or the Biblical statement is superfluous, which is not in harmony with the character of God, either. There are extreme cases in which people have said that they didn't need to study the Bible because God taught them directly — subjectively rather than objectively.[18]

Fourth, is the message true, that is, in harmony with the facts of reality? God doesn't bear false witness.[19]

Fifth, is the message significant? God never says trivial things.[20]

Sixth, has the "prophet" previously prophesied things which did not come to pass? God gives this as a clear test.[21]

Seventh, is it impossible to either prove or disprove whether the message is from God? Then we have no right to accept it since we are expected to test out everything.[22]

Eighth, does it glorify Jesus Christ? Here is the ultimate test of every spiritual thing![23]

The last "Mystical alternative" to discerning the will of God which we'll consider is the fatalist's easy way out: accept whatever happens with resignation, and "God's will be done!" But this just isn't true! Not everything that happens is pleasing to God!

—God isn't the only agent of change!
—God hasn't repealed the laws of nature!
—Not every occurrence *had* to happen! Words like "choice," "option," "alternative," "possibility," "coincidence," "might have been" and "should have been" have real meaning within the framework of the Biblical, supernaturalist world-view.

MY PLAN **AND** *GOD'S PLAN*

Having goals and working out plans for achieving those goals is hard work, but there's no escaping it! God expects us to know His will and to do something positive about it. God's will is a lot more specific and detailed than His plan. His plan leaves room for our initiative; His standards don't. His universal will provides the boundaries and the direction. As and when He has particular plans for us as individuals, the Lord of Heaven and earth is perfectly capable of doing something about it, and can be trusted to do so. Meanwhile, let's set our goals, plan our strategy, plot our course and work out our methods for doing battle against the mystical forces of Satan.

Are you experiencing a negative, emotional reaction against the idea of man-made plans and goals? Could it once again be that we have "tapped" the unnoticed influence of the slide toward mysticism? We will have to be objective about it and not allow ourselves to react subjectively and irrationally. Just remember four facts on which this principle is based:

First, because nature is orderly and predictable within the limits of our knowledge and our ability to exercise control over key events, the way the future will turn out, can be influenced.

Secondly, man is significant in history, an agent of change, designed by God to take an initiative in altering the future.

Thirdly, God has commissioned man to work together with Him toward His great plan for this universe.

Fourthly, opposition from mystical agencies is certain to be aroused, and must be overcome, just as surely as the complexities of mindless nature dare not be overlooked.

Both the spirit and letter of the Bible agree that people must make plans to bring about the will of God. Notice that the Apostle Paul made plans and set goals, some of which were either unsuccessful or needed to be altered. In the process of fulfilling his ministry, he had to take into consideration the other determining factors: God, Satan, other men and nature.[24]

Which Scripture passage first comes to mind when thinking about man making plans for the future? Probably, if you're like me, it's James 4:13-16. This is a warning to those individuals who would blithely affirm that they're going on an extended business trip, be gone for a specific length of time, and come out with a good profit at the end. However, you very well might not live that long, or circumstances might change drastically. We ought rather to qualify our statement by adding, "If the Lord wills, we shall live and also do this or that."

What does this Scripture actually teach? Does it tell us not to plan because God already has everything all pre-planned? Not at all! It instructs us to act in perfect harmony with the worldview which is the framework of all Scripture. It says to take the uncertainties of life and the will of God into consideration. Then, within these limitations we have every right and obligation to say, ". . . we *shall* do this or that!"

If it should happen that our plans can't be completed without going outside of God's will, then those plans must be changed.

Suppose for example that you set a goal and make a plan for attaining that goal. You say, "I'm going to become a competent Christian counselor, and in order to prepare myself, I'll enter the local university and study for the relevant degrees — if it's God's will." But suppose that after two years of study, it becomes obvious that the only way you will be allowed to continue on at that university is by acting as if you accept the principles of Secular Humanism, rejecting all objective standards of morality as well as believing that the theory of evolution fully accounts for man's origin. In theory and in actuality, you *could* continue on as planned — but not without going outside of God's will!

If totally uncontrollable and unexpected circumstances arise, well, that's the way it goes in this life. Not everything happens "on purpose." It could be that the Lord has decided to intervene. That possibility needs to be considered. But on the other hand, perhaps we should have prayed harder, worked harder, and planned more carefully. Certainly the man who sets out to manipulate the cause-and-effect events in nature can expect to be far more productive than the man who hasn't made any plans or set any goals. To do otherwise, is to disobey God's first commandment to man — to control the earth and rule over everything in it. Man has fallen — and we all often fail as a result — but never must we give up on becoming all that God has intended for us to be — responsible and efficient agents of change, working to glorify our Heavenly Father.

Notes — Chapter Fourteen

1. Matt. 11:29; Rom. 15:5; I Cor. 2:16; Phil. 2:5
2. Rom. 9:20, 11:33,34
3. This important spiritual principle is inspiringly illustrated and discussed in the little book:
   KUHN, Isobel: *Green Leaf in Drought-Time*; Chicago: Moody Press, 1957
4. Heb. 12:4-11
5. Acts 2:38,39; II Cor. 1:20-22; Gal. 4:6
6. For example: I Cor. 1:4,5; Eph. 1:15-23, 3:14-19; Phil. 1:3-11; Col. 1:9-12
7. I John 5:14
8. Matt. 12: 38,39
9. Judges 6:36-40
10. Heb. 1:1,2
11. Luke 1:18-20
12. I Thess. 5:20; II Pet. 1:19-21
13. Deut. 18:22; Matt. 5:18, 24:35; John 10:35; James 5:10
14. I Kings 22:21-23; Isa. 19:12-14; I Tim. 4:1; I John 4:1
15. Jer. 14:14; Ezek. 13:2,3
16. The Bible is not a book of "Delphic"-type oracles. The greatest single barrier to understanding is man's sinful predisposition and lack of effort. Amos 3:7,8; John 8:43-45; II Tim. 2:7; I John 5:20
17. Deut. 13:1-5; Psa. 111:7,8; John 5:31-39; Acts 10:43; II Pet. 1:19; Rev. 1:2
18. Any attitude or approach which nullifies the value and purpose of Scripture is inadmissible. Isa. 8:19,20; Luke 16:27-31
19. Num. 23:19; Titus 1:2; Heb. 6:18
20. Isa. 55:8-11; Jer. 33:3; Matt. 4:4; John 6:63; I Cor. 3:19
21. Deut. 18:22
22. I Cor. 2:15, 12:10; I Thess. 5:21; I John 4:1; Rev. 2:2
23. I Cor. 2:2, 3:11; Phil. 2:9; Col. 1:18, 3:4,11,17; I Pet. 4:11
24. Acts 16:6, 19:21, 20:13-16, 27:7ff.; Rom. 1:13, 15:22; I Cor. 4:19; II Cor. 1:15; I Thess. 2:18

15

Clearing Away The Mist

Praise God for showing the way out of confusion! Fallen man is so easily deceived. We are misled by preconceived notions and assumptions which have been passed on to us from others. Satan deceives us. Above all, we deceive ourselves by failing to "think our way out of it" — out of the self-centeredness, shallowness and inconsistencies of our own muddled minds. First, we ought to repent before God, confessing our sins of careless thinking. OK? Now let's bring forth some fruits worthy of repentance.

In this chapter, we will go through a number of different subjects which have lost their clarity in the mists of mystical musings. When we apply the world-view and conceptual thinking of Biblical supernaturalism, the clouds will begin to dissolve, the visibility improve, and our spirits rejoice!

JESUS AND HEAVEN ARE **REAL!**

Feelings may come and feelings may go, but the facts remain. While the psychical mind-set and it's wrong-side-of-the-tracks subdivision of mysticism fall into the sentient, feeling, subjective side; supernatural places and persons are objective facts which exist outside of us and are not at all dependent upon us — nor for that matter on this universe!

Jesus Christ and Heaven, His home, are the very essence of reality. Jesus is a real living, acting, thinking Person. Heaven is His environment, which is a temporary, disintegrating, unstable sort of place. He's there right now, but looking through the "window" at us.[1]

Haven't you often longed for Jesus to be more real to you? Does Heaven seem vague and intangible? Don't look inside yourself, because you won't find Heaven there. Stay out of the mists of mysticism! No internal experience can take the place of objective facts, even facts which we haven't as yet personally observed.

WORSHIP — IN SPIRIT AND IN TRUTH

Spiritual worship is God-directed. It looks out, away from the worshipper, not inward at the worshipper. There is only one legitimate basis for worship, and that is the truth — the facts — of who God is and what He's like. The more we know about God the more overwhelmed with awe, admiration, and praise we become.

Worship in spirit is the result of setting one's mind on the realm of the spiritual and the God who is Spirit.[2] Worship in truth concentrates on the rational and objective, not the irrational and subjective. In worship, we center our attention on Christ, not on self, sensation, or circumstances. A true devotional attitude toward the Lord and a "heart-felt" relationship with Jesus Christ are results which come from worshipping in spirit and truth.

Worship isn't an inward state of a man, or a mindless intensity of emotion, but the stark realization of God as He truly is. Worship, as every other spiritual thing, finds its meaning in Jesus Christ Who has revealed God's glory to us.

Counterfeit mystical devotion and worship can exist with little or no thought of God the Father or His Son, Jesus, even entering the mind. Worship which looks inward is man-centered. Worship which depends upon an artificial stimulation of some sort — a special environment, the sense of mysterious powers being present, continuous repetition of words or music — is sensual and of the world, and is mystical rather than supernatural.

MORALITY AND THE MYSTICAL DECEPTION

The world without God has no fixed moral standards. In all forms of Humanism, relativism rules the day. The dismaying thing is that Christians, slipping into subjectivist ways of thinking, are losing their hold on God's standards of holiness and purity. An example is the increasing number of Christian leaders who have fallen into flagrant sexual misconduct. We should be bitterly sad but not surprised. Sensuality and self-contained standards are built right into the mystical system.

The supernaturalist knows that Christian moral standards are clear, objective and unwavering. The supernatural, personal God has a character. Whatever is in harmony with His character is moral and good; whatever is less than, or contrary to His character, is immoral and sinful. We could rightly call Him "Ultimate Morality"! Jesus Christ came to display before our eyes the character and nature of His Father. The Scripture records it for us. God doesn't change, Jesus Christ doesn't change, His

Word never passes away. When founded on the Lord, moral standards don't change.[3]

Man changes. His "inner voice" doesn't always consistently say the same things. When moral standards are based on the subjective state of a person, they will be heavily influenced by circumstances, feelings, and internal enlightenings. All of us are tempted when our own lusts take over, so we're asking for trouble when we condition ourselves to respond to our internal drives and irrational impulses.[4]

The dangers of trusting in our *psyche* for moral guidance are are least three-fold:

—We're more likely to fall into sin when we've trained and conditioned ourselves to react to irrational and subjective impulses;

—Self rather than Christ, becomes the final standard of behavior;

—If we see God as doing His leading through a subjective, irrational moving in our spirits, our deceiving minds can justify anything, including gross sin, by claiming the leading of God. An example of this is the minister who left his wife and ran off with his secretary, and then insisted that this wickedness was by the direct leading of the Holy Spirit. Many of his followers accepted the explanation!

There are, theoretically at least, three possible ways to avoid the trap of moral relativism:

—God could personally reveal exactly what is right and what is wrong to each individual in each and every case. But He doesn't do it. "Conscience" is the closest that we come to this, and our consciences require training and sometimes make mistakes.

—We could have a Bible which gives us a specific commandment, either a "thou shalt" or a "thou shalt not," for each and every possible situation any person, in any culture, or any age, might find himself in. But we don't have that kind of Bible, and if we did it would be too big to carry around with us!

—We can do what God intended: we can study His character, nature and deeds, and then imitate Him. Make morality a rational choice rather than an irrational reaction. Apply principles and concepts drawn from His plainly-revealed "Personality Profile," the Scriptures, rather than get lost in the details of endless laws and rules. Saturate our minds so completely with Him that anything which is out of harmony with His character will be out of harmony with ours as well.

LEAVING GOD BEHIND

By absorbing the mystical way of thinking, the church is accomplishing exactly what Satan wants: drawing men's attention away from God and Heaven, and onto earth and mystical powers. Not only that, but it's happening so that the victims scarcely notice. Here's an example of how false teaching gradually takes over:

It is sometimes stated and often inferred, that there should be a *progression* of spiritual growth and commitment — from God to Jesus Christ and on to the Holy Spirit — and that anything other than this movement of attention is failing to "believe in the whole Trinity." The mystical trend is in the process of carrying the "progression" even beyond this. *In practice,* a form of pantheism is being produced. Pantheism is the belief in "god" as an all-pervasive, impersonal power or force which is working mystically throughout the universe. Of course, pantheism isn't being blatantly taught in so many words, but as a practice, it's very present and is quite obvious once pointed out.

Suppose the emphasis in our preaching and practice "progresses" from God *in Heaven* (the Father) — to God *with* us (the Son) — to God *in* us (the Holy Spirit). The next too-easy step is to move our emphasis on from "the *Holy Spirit*" — to "the power of the *Holy Spirit*" — to "the *Power* of the Holy Spirit" — to the "*Power . . .*" — and the transition into practical pantheism is complete. With time, when precedence has been given to the practice, a pantheistic definition of God must inevitably follow.[5]

Another pantheistic trap is the nice sounding but inaccurate cliche that "God is in all of us!" What must be asked is, "What do you mean by God?" Is God Some*one*, personal and transcendent, or is "God" an all-prevailing life-force which is contained within this natural universe? And in what way is God "in" all of us? And who is included in "all"?[6]

THE ROLE OF THE HOLY SPIRIT

There are few topics which evoke as much emotion, confusion and dissension in Christian circles as does the Holy Spirit. The blurring of the line between the supernatural and the mystical is nowhere more evident.

The Holy Spirit is real, just as God is real and His Son Jesus Christ is real. *Real* real! Supernaturally real. The Holy Spirit is more than a power, more than just the greatest of the spirits which are active in this universe, more than a — Someone — Who has taken up residence inside a Christian. The Holy Spirit is GOD, and can no more be limited to any particular location — for example, "inside" me as compared to "outside" me — than God can be contained and restricted either inside or outside this entire universe![7]

We don't "go on" from Christ to the Holy Spirit any more than we go on from God to Jesus Christ. Our only access to God is through Jesus Christ. To have Him is to have the Father; to honor Him is to honor the Father. It is Jesus who is the way, truth and life. Coming to Jesus is coming to God, and as a result the Holy Spirit comes to us. We are *Christians*, not "Godites" or "Holy Spiritists."[8]

The role of the Holy Spirit is not to lead our attention and interest away from Jesus Christ, but to make it possible to concentrate our minds on Jesus Christ. The Spirit comes from Christ, comes in the name of Christ, speaks of Christ and glorifies Christ. To have the Holy Spirit with us is to have the Spirit of Christ with us.[9] The Holy Spirit does not speak non-rationally because that would be contrary to the nature of God.

In short, the working of the Holy Spirit is not mystical, but is supernatural.

MIRACLES: SUPERNATURAL AND MYSTICAL

How shall we define a miracle? It may be too simple for some purposes, but I like the definition which says that a miracle is anything which would not have happened if nature had been allowed to run its uninterrupted course.[10] It is a miracle when God intervenes in the normal course of events. But there are other powers — mystical powers of mystical personages — which are also beyond the natural. Suppose something happens which is — or at least appears to be — impossible from the point-of-view of natural cause-and-effect. What should be our attitude? There are at least three possible explanations for such puzzling events:

—Perhaps it isn't really a miracle, and if we had enough information we could understand it;

—Perhaps it genuinely is a miracle from God; or,

—Perhaps it genuinely is unnatural, but is Satanic, mystical in origin, rather than supernatural.

As a consequence, miracles are not of much value for proving anything. For example, there was the university professor of philosophy who acknowledged the historicity of the resurrection of Jesus Christ from the dead, but who still didn't believe in the supernatural. After all, he said, it was just one of the many things which he couldn't explain.

The existence of mystical miracles explains the very real powers which can be found in both non-Christian religions and pseudo-Christian cults. We shouldn't be amazed at the hypnotic grip which some of the "way-out" leaders seem to have over their followers. Miracles and special powers may very well be present. It may also be possible that the "hypnotic" influence is demonic. Although people may do stupid things, they're not likely to be convinced by the mythical, but they easily may be by mystical magic. For this, the church itself must bear much guilt. The growing emphasis on subjectivism, and the failure to distinguish between the supernatural and the mystical, is paving the way for Christians in general and youth in particular to embrace the Eastern religions, cult groups or occultism.

THE INFLUENCE OF ORGANIZED MYSTICISM

As Christians, we have to live in the midst of societies which don't even claim to follow the same standards or to hold the same world-view as we do. The assumptions and attitudes which surround us are so intrusive that they're impossible to ignore, and difficult to oppose. Never forget that there is a general trend in the direction of mysticism. Remember also that traps don't look like traps, and the bait has to look good or it wouldn't catch the unwary. Be on guard.

Once we become conscious of the mystical influence and way of thinking, it crops up constantly. Notice the fad toward self-improvement courses which claim to draw out those hidden resources deep within each person. Sometimes the emphasis is on "unlimited human potential," or "the power of god which is available to all of us" — if we just unleash what is within. Perhaps they aim at helping a person to "free" himself from the restrictions of his conscious mind. The old sales motivation cliche, that "Whatever the Mind can Conceive and Believe, the Mind can Achieve," is another way of saying that whatever you can imagine subjectively, will become, by a mystical (divine?) power, objective reality. It's interesting to notice that when Paul said that he could "do all things through Christ" who strengthened him, he was referring to the ability to suffer deprivation and hardship![11]

Another characteristic of organized mysticism is the idolizing of human leadership. Almost all of the cults center around a strong and domineering personality who is usually believed to possess miraculous powers or unique authority. Whether it be a Jim Jones, an Eastern guru, a Joseph Smith or a radio or television preacher, it definitely is not Jesus Christ holding the reins of loyalty and affection.

TAKING UP THE SHIELD OF FAITH

There has never been a time when Christians needed protection from the flaming missiles of the evil one more than we do now. There is only one sure defense against the subtle and sometimes not-so-subtle influences of creeping mysticism: keeping Jesus Christ central in all things!

Every spiritual thing — belief, practice, method, attitude — must find its basic and essential meaning in the character, nature and deeds of Jesus Christ. Know Christ. Be totally absorbed in Him. Soak up every facet of His personality, His ways, His words, His actions. Make Jesus Christ the yardstick by which everything else is measured and evaluated. With the mind of Christ, we can think our way out of it!

Notes — Chapter Fifteen

1. Heb. 12:1,2
2. Isa. 6:1-5; John 4:21-24; Rev. 4:9-11, 15:4
3. Rom. 1:24-32; Eph. 5:1-6; Col. 3:5-10; I Thess. 4:1-8
4. Heb. 5:14; James 1:13,14; II Pet. 2:14
5. The transition has already been made in more "liberal" circles. Setiloane (1986: p.20) utilizes the term "Ultimate Reality" to describe his animistic "Vital Force." He borrows the expression for Bishop John A.T. Robinson, whose pantheism helped launch the so-called "God is dead" movement. Setiloane (p.39) sarcastically wonders why Protestants haven't seen how close this "Vital Force" is to the Christian idea of the Holy Spirit. Some eventually will.

    ROBINSON, John A.T.: *Honest to God*; Philadelphia: The Westminster Press, 1963

6. As "Christians" become more engrossed in mystical thinking, the trend will not only be toward pantheism but inevitably toward universalism as well. The Gospel is replaced by union with an All-prevailing force, and those who are in union with this Force — that is, exhibit evidence of the correct subjective experiences — will be accepted on that basis. Thus mysticism nullifies the historic facts of the Gospel.
7. I Kings 8:27; Psa. 139:7-12; Acts 5:3,4
8. John 1:18, 14:6-9; Col. 1:18; I Tim. 2:5; I John 2:23, 5:12
9. John 14:16-18,26, 15:26, 16:7,13-15; Rom. 8:9-11
10. C.S. Lewis (1947: p.10) uses this simple but practical definition in his little book, *Miracles*.
11. Phil, 4:13

16

Doing The Truth

The great advantage which truth has over falsehood is that it doesn't clash with the facts! Beliefs must be applied to real situations, so believing the truth is important. The restored Biblical world-view was true, so it actually worked when put into practice, producing the scientific and technological age in which we live. The same supernaturalist way of thinking proves itself as both superior and true in everyday life situations. After clearing up our thinking, we can live the practical lives of supernaturalists in this natural world in which God has placed us.

We say this with the realization that psychical powers are also real. They also "work." No one can deny that the purely psychological workings of the human mind — our mental and nervous conditions — have a great influence on our health and abilities. In addition to this, the reality of mystical forces and beings must not be ignored. However, the results produced by such powers as these won't help us fulfill the commission given us by God. Strengthening the psychical is no way to promote the supernatural!

The superiority of the supernaturalist mind-set in meeting the needs and challenges which face the church today is evident in each of the following areas of practical Christian concern:

THE CHRISTIAN AND COUNSELLING

It's been said that psychiatrists are the "witchdoctors" of the modern twentieth-century world. One survey listed them as holding the highest status-position of any profession in the United States. It's also sadly significant to learn that psychiatrists were found to have the highest suicide rate among thirteen major professions. In my opinion, there are legitimate grounds for despair among psychiatrists who sincerely want to help people, but have no other tools than those supplied by humanism. Humanism isn't true. It leaves out God and man's God-given nature. It can't be expected to be effective because it doesn't agree with reality.

Recently my wife and I had a discussion with a clinical psychologist who stated that she was aware that she had been leaving out the "spiritual element" in her therapy. As a consequence she was looking into the Eastern religions. By spiritual, of course, she meant mystical. The idea of a supernatural God was totally abhorrent.

As the world — and with it, the church — drifts further toward mysticism, the counselling problems will be compounded. People who are conditioned to respond to their emotions, their intuitions, their urges, become enslaved to irrational impulses: lack of discipline, unstable behavior, roller-coaster emotions, weakened values and standards, self-centeredness.

In addition to the psychological hang-ups which the slide toward mysticism leaves in its wake, the counselor dare not forget the very real presence of demonic powers. There will be times when the only explanation for a counselee's problems is in demonic affliction, and the only solution is to face up to the powers of darkness. In an age of neo-paganism, mind-altering drugs, and open occultism, it ought to be no surprise. By the way, the counselor doesn't need some special mystical power within to deal with demons. The authority of the supernatural Son of God, as available to every Christian, is sufficient.

Christian supernaturalism has a world-view that works in counselling, which neither naturalism nor mysticism can equal. What could be more realistic and practical than the "Law of the Harvest" — that what a man sows is what he reaps.[1] A Christian counselor ought to consistently apply the principle of cause-and-effect: do the right things and you can expect to get better results than when you fail to do the right things. Make things happen. Think your way out of it! Disorganized lives can be re-organized. Tangled, knotted strings of events can be patiently and methodically straightened out and unsnarled. Platitudes? No! A workable counselling philosophy? Yes.

While rational, sensible, Godly living does wonders for messed-up lives, the task of building new habits of thinking and behaving is hard work. The only way a discouraged and hurting person is likely to gain and maintain sufficient motivation to see it through, is by taking on a new world-view. When a person realizes that the Lord is real and active, that he himself is an agent of change in a world which has a framework of natural cause-and-effect, he is no longer trapped and hopeless, a victim of circumstances or forces which he can neither understand nor control. He prays to a real God, enjoys real freedom from guilt as a result of the doing and dying of a real Savior, and is free to take an active part in his own rehabilitation. Praise God, there is a way out![2]

THE CHRISTIAN AND WORLD MISSION

The popularity and growth of mysticism in the West is no more than a return to what most of the world has never left. Every Christian should be prepared to share his faith, and consequently, needs to be aware of the implications of mysticism to the evangelizing of the world.

The example of a Christ-like life includes living according to a supernaturalist's world-view, but as vital as this is, an example isn't sufficient to communicate that world-view. Observing *what* we do doesn't explain *why* we do it. Teaching which begins at the beginning is essential. It isn't by accident that Genesis 1, with its world-view, comes before John 3, with the Gospel.

Christian, supernaturalistic thinking isn't even a viable alternative to a mind saturated with mystical assumptions. The Christian way of reasoning is nonsense, Christian values are irrelevant. Arguments won't work until attitudes change.

Right attitudes won't come without understanding facts previously unknown. Understanding comes with a change in word-view. Pre-evangelism — preparing people to listen to the Gospel — is more imperative than ever. People must be led into a true knowledge of Jesus Christ,[3] but effective communication will take place only when we take into consideration the belief-system of our listeners. They are not likely to understand us if we can not understand *them*. Often we fail to understand ourselves!

Another danger in forgetting that the world thinks mystically, is the risk of producing a Christianized form of heathenism. This is what happens when Christian terms, practices, and institutions are plugged into the framework of a mystical mindset. On the outside, it might look Christian, but the similarities are likely to be more superficial than actual. When crisis comes, so does reversion to old ways.

Cross-culturally, this presentss a special danger. The modern missionary is constantly reminded that his obligation is to "contextualize" the Gospel; to evangelize, and not to Westernize. The Gospel must be separated from our culture and allowed to find its natural expression within each new culture to which it is introduced. The problem is, that to mysticism in general, and most especially in animistic cultures, *everything* has "religious" significance. Almost nothing is harmless and neutral from a spiritual view-point.

Rather than overdo our guilt at allowing our cultural background to influence those of other cultures to whom we preach, we ought to recall the extent to which our thought-structure has been, and still is, being molded by Christian presuppositions. Even with the "neo-paganism" of the West, the influence of a Biblical world-view and style of reasoning hasn't been discarded entirely. Simply because some characteristic is well-entrenched in the West does not make it non-Christian. Just as the hidden connotations of a non-Western culture may be rooted in mysticism, so also, the hidden connotations of Western culture may be rooted in Christian principles.

The growth of churches, and the development of a strong and competent Christian leadership which handles the Word of God properly, both require a solid foundation in supernaturalism, not mysticism. Without the true Biblical world-view, the Scriptures can not be understood. Without the conceptual thinking style which is built into the Word, the application of Biblical principles and precepts can not be made.

When the church wakes up to the threat of the mystical

invasion, the response ought to be dramatic. The realization of the danger of sliding away from a Biblical faith, of the falseness of non-Christian and pseudo-Christian religions, and the renewed sense of urgency and compassion for the lost — all seek to drive and empower to preach Christ!

THE CHRISTIAN AND SCIENCE

The Christian who continues to hold to the Biblical worldview has no problem with science as such, because he has no problem with nature. To him, nature is indeed nature, the reasonable system created by the reasonable God. His argument is not against science, but against a view which sees nature as being a system closed to influence from God.

The Christian has a better foundation for science than either the naturalist or the mystic. The naturalist can't really account for his own rationality — thus his drift toward mysticism. The mystic doesn't see the natural system as all that significant. Mystical powers, not scientific explanations, occupy his interests. I don't know whether scientists will ever make good mystics, but mystics will never make good scientists.

The sciences, in common with all other fields, have more non-Christians in them than Believers. That doesn't necessarily make science anti-Christian, or Christians anti-scientific. Our modern world is dominated by those who have control over the sciences, technology and economy, and pitifully few of them are Christians. Yet the origin of the international power structure had its roots in the Christian faith of the earlier scientists and thinkers. Perhaps the church should set out to win back the territory which it deserted to the opposition. If non-believing scientists embrace the modern mystical fad, who else but the Christian is left to deal with nature in the way God intended?

THE CHRISTIAN AND BUSINESS

The business community is seldom accused of being overly ethical! That doesn't reflect badly against business, only against sinful human nature. The Christian need not be at a disadvantage in the business world. If the believer in business is militated against for being honest, he's trusted for the same reason. Society should only benefit if control of the economy were in the hands of Godly individuals. Assuming spiritual motives, the Christian has no cause for feeling guilty over making a profit, building an industry, or employing workers. In fact, there are specific reasons for expecting the Christian, because of the supernaturalistic outlook, to perform with above-average efficiency as a business administrator or manager.

The Christian business person ought to possess a strong sense of responsibility to be productive, accountable and a good steward of both material and opportunity. The supernaturalistic world-view encourages consciousness of the fact that actions have consequences. Man is an agent of change, capable of planning and setting goals. The principles of cause-and-effect apply to the Christian in business in the same way that they do in everything else.

The supernaturalist's ethical base for loyalty and commitment is greater than the non-believer. Neither is there an unrealistic belief that God is obligated to cause financial prosperity and success without hard work and wise business practices.[4]

Another advantage of the supernaturalistic world-view is that it encourages a thinking-style with a priority on the conceptual. Getting lost in the details is a sure route to failure as a manager or administrator.

THE CHRISTIAN AND EDUCATION

Education is another field being deserted by Christians and

claimed by Humanists. If we can succeed in casting off the influence of mystical attitudes, we will stop degrading our intellects, give up idolizing irrational "knowledge," and become more responsible in the use of our God-given brains. Since the truth always works best, and since the Biblical world-view is the truth, Christians ought to have a distinctive advantage when it comes to thinking.

Christians ought also to be deeply concerned with the influence of mysticism within the public educational systems, but to the degree that mysticism is in the churches it will also be in Christian schools. As the world goes ever more mystical, the schools may logically be expected to lead the way. Where else should the battle for the mind take place? If we fail to resist, how will we justify our negligence before God?

THE CHRISTIAN AND HEALTH

No one hates his own flesh. In fact, preoccupation with health has almost kept pace with the extent to which we neglect our health! The supernaturalistic world-view has a big practical advantage when it comes to stewardship of our bodies, and most of us sorely need it.

If we recognize the inevitable influence of natural cause-and-effect, we will be more likely to think of the consequences of how we treat the physical marvel we temporarily occupy. If we don't take care of it, we will inevitably pay the price.

Because we know that every individual is subject to the same whims of nature as anyone else, we will not go on a guilt-trip every time we catch a virus or some other bug. Rather than assuming a mystical explanation — a lack of faith or an attack by evil forces — we'll give first option to a normal, impersonal nature.

Another real big advantage, which might not look very

important, is that the supernaturalist doesn't feel obligated to hate his own body as "fleshly." He knows that God intends to save us from being disembodied spirits! He doesn't need to fall into what the philosophers call the "mind-body dichotomy." He can be a whole person before God.

THE CHRISTIAN AND HIS ENVIRONMENT

Ever since the Fall, man has been engaged in a battle for survival with his natural surroundings. The supernaturalist's world-view, even in that version which has been appropriated and mutilated by naturalism, has saved mankind from untold suffering. In those sections of the world where the concept of natural cause-and-effect is most ingrained, man has the technology and expertise to protect himself against exposure, starvation and a great many diseases and health hazards.

It has been said that Christianity is the only "naturalistic religion" (not meaning naturalistic in the sense that we've used it, a belief in a closed system of nature which has no influence from God or the supernatural) but rather as a belief-system which sees nature as truly natural. Thus, *supernaturalist* Christians can be practical and meet the real needs of the real world.

Of all people, the Christian ought to love and cherish his natural environment. It reflects the glory of its Creator. Nature is neither his enemy nor his master. He can appreciate it for what it is, a system of things which God made. Why he made some of them, like mosquitoes and lice, I don't know, but presumably they've degenerated from some originally useful functions. God made everything beautiful in its day — even if perhaps that day has come and gone for some. At any rate, even the little things that creep, hop and buzz around, have a beauty all their own.

From the supernaturalistic view-point, man has a clear

responsibility to take care of the environment. God will hold him accountable for what he does with his world. His responsibility is that of caretaker over an unreasoning, mindless but wonderfully beautiful system. Nature is helpless. It can't do anything but just be what God made it. As God's representatives, His agents of change, we can do something with it. As it stands right now, there's room for improvement, mostly because of past human irresponsibilities. As Christians with a supernaturalist world-view, we will use it wisely, protect and, as much as we can, restore it. But we will neither worship nor abuse it.

Godless exploitation and careless use of the earth's materials have resulted in the well-publicized environmental hazards of pollution, contamination, erosion, acid rain, the "ozone hole," endangered species and depleted resources. Because of his selfishness and disobedience to God's commission, man hasn't ruled — he's raided! He's demolished rather than domesticated.

No less destructive and disobedient are those societies which have traditionally been the deepest into mysticism, the very ones who have worshiped and served the creature rather than the Creator. They also are the ones which are hurt the most by nature. It is the part of the world without a Christian heritage which suffers the most, and the reason is mysticism. Mysticism, in spite of its philosophy of harmonizing with the environment, can not produce the evidence to back up its claims. It isn't true to the facts, and simply doesn't work! Animism, which is mysticism in its most basic form, has allowed nature to ravage, destroy or restrict those who are held in its trap. "Christian" mystics, ignoring, or even denying the influence of natural cause-and-effect on their lives, are not in a much stronger position. No one disregards nature without suffering the consequences. The claim of freedom from sickness and poverty, because these are all direct acts of Satan, is not only inaccurate as a concept, but doesn't work in practice.

Part of the modern and Western mystical philosophy is the concept that the earth must be viewed "holistically" — as a single whole rather than as separate components — and "ecologically" — more-or-less meaning that all components are interdependent. A common conclusion is that the earth, itself, is a living organism of which each person is an infinitesimal segment. Man is only part of the system, rather than ruler over the system. If followed to its ultimate logic, the most primitive hunters and gatherers must have been the most successful at being a part of the environment. Care to try it as a life-style? It isn't very good as a modern survival technique, considering the big population and shrinking wilderness. The fact of the matter is that it never has worked very successfully.

A society which blends too closely with its environment leads an animal-like existence, maintains only a subsistence level in its standard of living, and is at the mercy — more accurately, the lack of mercy — of drought, disease or any other natural disaster. Thus if man becomes too "ecological" in his life-style, nature rules him rather than the reverse. I don't think we want to be reduced to the level of primitive animists.

THE CHRISTIAN AND SOCIETY

The Christian supernaturalist not only approaches nature as something truly natural, but treats man as truly man. It is not possible to hold to the supernaturalistic world-view without having a social conscience. Society has value because, however low it has fallen, society is still composed of people: sentient, self-conscious and significant, the objects of God's love and compassion. Those who have absorbed even a little of the character, nature and deeds of God can't keep from caring about people.

However, evangelical Christians are commonly accused of being unconcerned and uninvolved with social issues. Is it true? Sometimes, in some ways, the criticism is justified. Could the current lapse of social concern be due to the mystical influence in the church? Historically it has been Christians who have built the hospitals, set up the orphanages and homes for the elderly, founded the greatest universities and sent out medical missionaries. Over the years, most social reform has originated with Christians. And today? If there are areas where Christians are not involved, it might just be in those places where Humanistic principles are being promoted under the guise of Christianity, in which case no true child of God wants to be involved. And it just might be, that there is also enough mysticism in the church to draw Christians away from the real world.

A supernaturalist's outlook toward nature and toward himself, as one commissioned by God to actively initiate changes in the world, will make him more of an asset and less of a burden to society as a whole. He will be more responsible and honest in his job, less often found in the divorce courts, more law-abiding, less racist, more giving and less taking, more loving, and less shoving.

The world can not afford to lose the benefits of Christian compassion and social concern! Society is hurting. People are in pain. As a consequence of working and counselling with many people, from many national backgrounds, races and cultures, in many different situations, I've personally come to the point of assuming that each new acquaintance is struggling with a heavy burden of one nature or another, and am rarely proved wrong. This is realism, not cynicism. In the most affluent societies in the world, in the ones which are turning away from their Christian heritages, people-problems are increasing as the populations grow: destitution, poverty, illiteracy, hunger, disease, anger, fear, hatred, guilt, shame, insecurity, anxieties, alienation, loneliness, perversion, neuroses and violence. That's the good news; elsewhere it's worse! Is there any society in the world which doesn't desperately need people who care enough to do something that will help? Mankind simply isn't coping.

The Christian, if he is free from mystical influence, will be genuinely concerned about the physical and material needs of people. Having one's eyes set on Heaven doesn't make one oblivious to earth! Not always is an individual's sorry predicament his own fault, and often, even when it is, there is little he can now do about it unless someone is willing to supply some practical assistance.

The mystical mind-set is not designed to solve practical social problems. Ask the sick person who has picked up a communicable disease and been told that he doesn't have enough faith. Ask the fifteen-hundred or more beggars who will die on the streets of Calcutta next year. It simply isn't a fact that if people were to get in tune with the mystical powers, their physical and material problems will either spontaneously solve themselves or else become trivial. Both natural and psychological cause-and-effect are still in operation. Something realistic must be done.

Christians, like Christ, have a mission in the real world. Do we act like missionaries? Does the world see us as an active force in society, or as a people all wrapped up in themselves? Can they see that we believe ourselves to be God's agents of change, carrying out a commission to change the direction of history? Or, do they still have the impression that there are activities with invisible forces and the inner workings of men? Are we psychics or soldiers?

Concern about practical issues, socially responsible behavior, a stable personal life, a conscientious and unselfish use of money and material possessions, and a realistic outlook toward the cause and cure of social problems — these ought to be strongly characteristic of every Christian. Throughout history, the disintegration of a society has taken place when the church has become introverted, and abandoned its role as the Salt of the Earth.

Jesus wasn't a mystic. He didn't lead an introspective life, nor live in a dream world, nor overlook one aspect of a man's nature while ministering to another. He went about doing good. His miracles were to satisfy needs rather than curiosity. Jesus was eminently practical. So ought to be His people.

Notes — Chapter Sixteen

1. Gen. 4:6,7; Luke 6:38; II Cor. 9:6; Gal. 6:7-9
2. The titles given to certain theories of counselling illustrate the concept that man is not a victim of circumstances over which he has no control. E.G. "Reality Therapy," "Integrity Therapy," "Neuthetic Counselling."
3. I Tim. 2:4; II Tim. 2:25; II Pet. 1:3,8
4. Much has been written regarding the "Protestant work ethic," and not always favorably in this modern climate of academic disapproval of capitalism. Whether one disapproves or not, and in spite of abuses and unethical practices, there is clearly a connection between Protestantism and the ethic of hard work and productivity as a Christian obligation.

17

Where Is It All Going To End?

We lift our eyes from the many little personal details which occupy our minds. We look out and around the world at peoples and nations, philosophies and institutions. We see great trends changing the world! Ideas and systems are vying for supremacy! Humanists vow to liberate the world from "supernaturalistic religions." Respected Christians warn us against the threat of the new mysticism. Scandals besmirch the public image of the church. Almost everything seems to be in a state of flux. Where is history heading?

Remember that "hole," the door between Heaven and earth which we spoke about many chapters ago? One of these days it's going to open up again, and get bigger — and bigger — and

BIGGER! Just imagine it: the sky will roll up like a scroll! The SKY! Space is going to split wide apart between this natural universe and God's Heavenly Realm![1]

This isn't something figurative and symbolic we're talking about. This is history — future history. It will happen just as surely as the death of Jesus Christ was a real event in history, just as surely as He was buried in a hole cut in a little rocky hill on the north edge of Jerusalem; history as real as His coming back to life again with a body of flesh and bone. This is objective reality that we're talking about, not the kind of "reality" which takes place somewhere inside a person. Human eyes will be able to see it take place. Fingers will point, traffic will snarl and buildings will empty into the streets. Ecstacy and terror, joy and fury, laughter and wailing.

Through that mammoth rent in space will come riding a Person — a fantastic, glorious, powerful, superhuman Person. Not a ghost or a vision, but a real, living Being — our Lord.[2]

And He'll be riding a horse! A white horse. Don't ask me where the horse came from, or if it's a literal horse, or something that comes closer to looking like a horse than anything we're familiar with. But as for myself, I find the idea of a horse more suitable — and vastly more aesthetically pleasing! — than a space ship, or some kind of spectral image in the air, or than spiritualizing the whole thing away as only symbolism. The details might be a bit uncertain, but the event is literally and actually going to happen. And I'm sure that if He's decided to ride a horse, He's capable of getting one!

This isn't any "Gentle Jesus, meek and mild," but a King and a Soldier. God will have reached the end of His patience! When things have gone as far as He will tolerate, His glorified Son is personally entering the war. And that will take care of the opposition!

He won't be coming out of that gap in space, alone. At this very moment — except that their "moments" are totally independent of ours — His army might be mounting their own horses, pawing the Heavenly turf, impatient to charge out of that Universe and into this one. Won't that be something! It will be the cavalry charge to end all cavalry charges! His mighty angels, Heaven's warriors, as powerful as fire-storms, will plunge into battle.[3]

And there will be others with Him as well — the spirits of just men made perfect, the dead in Christ, the saints of the ages — Christians just like us, who have been waiting eagerly to join us again. For us, that will be the moment when The Big Change takes place and we get our new bodies, the ones which will be constructed of the same kind of substance as His, super and indestructible. At that precise moment, "eternal life" will be ours by experience rather than by faith.[4]

Right now, at this very moment in the present, precious people are losing their hold on reality and sliding into the mists of mysticism. A whole universe — God's Universe — is being lost from their world-view. When they grasp the reality of Jesus Christ, they won't feel the need to look elsewhere. Let's make a plan to do something about it.

Notes — Chapter Seventeen

1. Isa. 34:4; Rev. 6:14
2. Rev. 19:11ff.
3. Matt. 13:41, 16:27, 24:29-31; II Thess. 1:7; Heb. 1:7
4. I Cor. 15:50-58; I Thess. 4:13-18; Heb. 12:23

Bibliography

Barclay, Oliver, *The Intellect and Beyond: Developing a Christian Mind.* Grand Rapids: Zondervan, 1985.

Barcus, Nancy. *Developing A Christian Mind.* Downers Grove, IL.: Inter-Varsity Press, 1977.

Book, W. David. ed. *Opening the American Mind: The Integration of Biblical Truth in the Curriculum of the University.* Grand Rapids: Baker, 1991.

Blamires, Harry. *The Christian Mind.* London: Spok, 1963.

_____. *Recovering the Christian Mind.* Downers Grove, 1969.

Clark, David, and Geisler, Norman. *Apologetics in the New Age: A Christian Critique of Pantheism.* Grand Rapids: Baker, 1989.

Corduan, Winfried, and Geisler, Norman. *Philosophy of Religion.* Grand Rapids: Baker, 1988.

Chapman, Colin. *The Case for Christianity.* Grand Rapids: Eerdmans, 1981.

Dyrness, William. *Christian Apologetics in a World Community.* Downers Grove: Inter-Varsity, 1984.

Evans, C. Stephen. *Existentialism: The Philosophy of Despair and the Guest for Hope.* Grand Rapids: Zondervan, 1984.

_____. *Philosophy of Religion.* Downers Grove: Inter-Varsity, 1985.

Geisler, Norman. *Is Man the Measure? An Evaluation of Contemporary Humanism.* Grand Rapids: Baker.

Geisler, Norman, and Brooks, Ron. *When Skeptics Ask.* Wheaton, IL: Victor, 1990.

Geisler, Norman and Feinberg, Paul. *Introduction to Philosophy: A Christian Perspective.* Grand Rapids: Baker, 1987.

Geisler, Norman and Watkins, William. *Worlds Apart: A Handbook on World Views.* Grand Rapids: Baker, 1989.

Gill, David W. *The Opening of the Christian Mind.* Downers Grove: Inter-varsity, 1989.

Grodthuis, Douglas. *Unmasking the New Age.* Downers Grove, IL: Inter-Varsity, 1986.

Guiness, Os. *The Dust of Death.* Downers Grove: Inter-Varsity, 1974.

Hasker, William. *Metaphysics, Constructing a World View.* Downers Grove: Inter-Varsity, 1987.

Henry, Carl F.H. *Twilight of a Great Civilization.* Westchester, IL: Crossway, 1988.

Holmes, Arthur. *Contours of a World View.* Grand Rapids: Eerdmans, 1986.

_____. *The Making of a Christian Mind.* Downers Grove: Inter-Varsity, 1985.

Hoffecker, W.A. and Scott, Gary, eds. *Building A Christian World View*. Vol. 1: *God, Man, and Knowledge*. Grand Rapids: Baker, 1986.
_____. eds. Vol. 2: *The Universe, Society and Ethics*. Baker, 1988.
Kreeft, Peter. *The Best Things in Life*. Downers Grove: Inter-Varsity, 1984.
_____. *Socrates Meet Jesus*. Downers Grove. Inter-Varsity, 1987.
Macaulay, Susan S. *How to be Your Own Selfish Pig*. Elgin, IL: D.C. Cook, 1982.
Moore, Peter C. *Disarming the Secular Gods*. Downers Grove: Inter-Varsity, 1989.
Nash, Ronald. *Faith and Reason; Searching for a Rational Faith*. Grand Rapids: Zondervan, 1988.
Newport, John P. *Life's Ultimate Questions: A Contemporary Philosophy of Religion*. Dallas: Word, 1989.
Peterson, Michael L. *Philosophy of Education*. Downers Grove: Inter-Varsity, 1986.
Pinnock, Clark. *Reason Enough*. Downers Grove: Inter-Varsity, 1980.
Ratzsch, Del. *Philosophy of Science*. Downers Grove: Inter-Varsity, 1986.
Schaeffer, Francis A. *The God Who Is There*. Downers Grove: Inter-Varsity, 1986.
_____. *He Is There and He Is Not Silent*. Downers Grove: Inter-Varsity, 1972.
_____. *How Should We Then Live?* Old Tappan, N.J.: Revell, 1976.
Schlossberg, Herbert, and Olasky, Marvin. *Turning Point: A Christian World View Declaration*. Westchester, IL: Inter-Varsity, 1990.

Sire, James W. *Discipling the Mind*. Downers Grove. Inter-Varsity, 1990.

———. *How To Read Slowly: A Christian Guide to Reading With The Mind*. Wheaton: Harold Shaw, 1989.

———. *The Universe Next Door*. 2nd ed. Downers Grove: Inter-Varsity, 1988.

Stott, John R.W. *Your Mind Matters*. Downers Grove: Inter-Varsity, 1972.

Timmerman, John, and Hettinga, Donald. *Reading and Writing as a Christian*. Grand Rapids: Baker, 1987.

Thiselton, Anthony. *The Two Horizons*. Grand Rapids: Eerdmans, 1980.

Trueblood, Elton. *Philosophy of Religion*. Grand Rapids: Baker, 1976.

Veith, Gene, Jr. *Reading Between the Lines. State of the Arts: From Bezalel to Mapplethorpe.*

Wilson, Douglas, *Recovering the Lost Tools of Learning.*

Van Leeuwen, Mary S. *Christianity and Psychology*. Grand Rapids: Eerdmans, 1987.

Walsh, Brian & Middleton, Richard. *The Transforming Vision*. Downers Grove: Inter-Varsity, 1984.

Walters, Albert M. *Creation Regained*. Grand Rapids: Eerdmans, 1987.

Wolfe, David. *Epistemology*. Downers Grove: Inter-Varsity, 1982.

Woltestorff, Nicholas. *Reason Within the Bounds of Religion*. Grand Rapids. Eerdmans, 1984.

Yandell, Keith E. *Christianity and Contemporary Philosophy*. Grand Rapids: Eerdmans, 1986.

Through the Eyes of Faith Series.
San Francisco: Harper & Row.

Chewning, R.C., Eby, J., and Roels, S. *Business Through the Eyes of Faith.*

Lundin, Roger and Gallagher, Susan. *Literature Through the Eyes of Faith.*

Myers, D.G. and Jeeves, Malcolm. *Psychology Through the Eyes of Faith.*

Wells, Ronald A., *History Through the Eyes of Faith.*

Wright, Richard T. *Biology Through the Eyes of Faith.*

The Turning Point Christian World View Series.
Westchester, IL: Crossway

Bandow, Doug. *Beyond Good Intentions: A Biblical View of Politics.*

Beisner, E. Calvin. *Prospects for Growth and Prosperity and Poverty. The Compassionate Use of Resources in a World Security.*

Billinsley, Lloyd. *The Seductive Image: A Christian Critique of the World of Film.*

Curry, Dean C. *A World Without Tyranny: Christian Faith and International Politics.*

Myers, Kenneth A. *All God's Children and Blue Suede Shoes.*

Olasky, Marvin, and Olasky Susan. *More Than Kindness.*

Olasky, M., Schlossberg, H., Bethoud, P., and Pinnock, C., *Freedom Justice and Hope: Toward a Strategy for the Poor and Oppressed.*